Virtual Training for the Clueless®

How to use the latest tools of technology to help you make money, save money, and save time.

Mike Rounds

Virtual Training for the Clueless®
Paperback – Pubished 2014

ISBN: 978-1-891440-65-6

This book is written with the understanding that the author was not engaged in rendering legal services. The information included has been carefully prepared and is correct to the best of his knowledge as of the publication date. If you require legal or expert advice, the services of professionals should be used. The author disclaims any personal liability, either directly or indirectly, for advice or information presented in this book.

The information as described herein has been used successfully to obtain profitable business for some of the persons who have used it. Although all efforts have been expended to supply the latest in complete, accurate and up-to-date information, it must be understood that the ultimate success of the user is dependent upon market conditions, efforts expended by the user, and other variable factors that are beyond the control of the authors, and that neither the users' actual expenses nor profits are guaranteed nor implied.

Throughout this book, trademarked names are used. Rather than put a trademark symbol after every occurrence of the trademarked name, we used the names in an editorial fashion only, and to the benefit of the trademark owner, with no intention of infringement of the trademark.

At the time this edition was printed and released, all of the sites listed herein were active and accessible to anyone having access to the Internet. Neither the author nor the publisher is responsible for broken links, abandoned sites, or changes that are beyond their control.

Books may be purchased in quantity and/or special sales by contacting the publisher, CPM Systems, at 6318 Ridgepath Court, Rancho Palos Verdes, CA 90275-3248; 310-544-9502 or by e-mail at CPMPub@RoundsMiller.com.

CPM
SYSTEMS
PUBLISHING COMPANY
6318 Ridgepath Court
Rancho Palos Verdes, CA 90275-3248

Printed in the United States of America

Contents

Introduction

The first time someone sent instructions for doing something using the telegraph, we entered the world of virtual training and presentations.

Of course, we've come a long way since then with audio, video, interactive training programs, and mobile computing; the virtual world has literally exploded.

Introduced during the late 1980's and refined for everyday usage in last five years, virtual delivery of training and presentation materials has now become an available technology for anyone who can perform basic computer skills like word processing, PowerPoint generation, and narrating background material for slides.

The integration of digital cameras and video recording into smart phones, coupled with on-line resources like YouTube®, have demonstrated to the world that whatever you need to record, you can, both simply and inexpensively, and share it with the world instantaneously.

Some companies are ahead of the curve with use of virtual presentations, and they are making it a strategic advantage.

Some are behind the curve and they're being dragged kicking and screaming to use it.

However, the reality is, whether you love it to hate it, **it's here to stay.** The question you need to ask yourself: If it isn't your competitive advantage, how are you going to compete with people who are making it their competitive advantage?

If you've wanted to use virtual trainings and presentations to profitably deliver your seminar and training sessions, watched others profiting from their virtual programs and said to yourself *I really need to do that for my programs, or* think it's too complicated, too expensive, or just beyond your technical skills, you're about to learn that it's simple, inexpensive and highly lucrative.

Welcome to the world of Virtual Content Delivery!

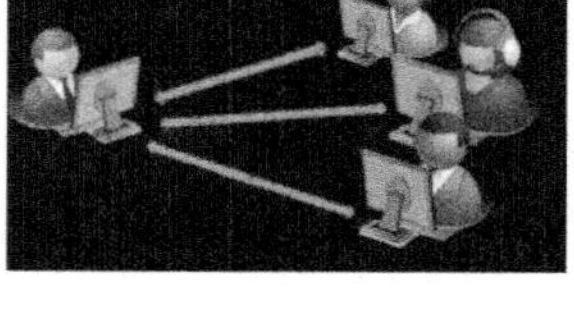

Part 1 - The Basics

What is virtual content delivery?

The basic concept of virtual content delivery is replacing the face to face transfer of information with technological tools such as radio, television, and more specifically, the web.

What is Instructor led Training? (ILT)

Instructor-Led Training (ILT) refers to training that is delivered in person to an audience that is the same physical proximity and the person delivering the information.

What is Virtual Instructor-Led Training (VILT)

Virtual Instructor-Led Training (VILT) refers to training or any type of presentation that is delivered in a virtual or simulated environment, or when instructor and learner are in separate locations.

How is VILT used and why use or even consider using VILT?

In the world of virtual content delivery a virtual lecture hall is an online learning environment where the environment can be web-based and accessed using a computer system with web access.

Just like in a real-world lecture hall, a participant in a virtual lecture hall may participate in synchronous instruction, which means that the instructor and participants are logged into the virtual learning environment at the same time.

Virtual lecture hall software applications often employ multiple technologies, such as telephone communications, web conferencing, video conferencing, live streaming, and web-based VoIP to provide remote participants with the ability to collaborate in real time. Some software applications may also provide real-time communication tools, such as message boards and chat capabilities.

Businesses are employing it as part of broader blended solutions for training and development while other common business-related uses include customer and product training, professional development, and sales training.

Simultaneously, many colleges and universities offer established online graduate and undergraduate degree programs based on VILT technology.

The design of VILT programs requires a different approach than classroom or online asynchronous programs. Learning content must be developed with maximum student engagement in mind.

The Different Uses for Virtual Content Delivery

This is going to be one of those areas where three people will have a dozen differing opinions as to what you should use virtual training for and why using a virtual delivery system is advantageous or dis-advantageous for a particular application.

Let's start by saying that virtual presentations and trainings are replacing face-to-face communications with the tools of communications technology and that the use of VILT must be appropriate for your needs to be successful.

The word appropriate, loosely defined, means correct and proper.

Now the question is: appropriate for whom, for what topics, what audience, and for whose available budgets of time and money?

In other words virtual technology isn't a panacea for the world's information delivery problems! On the other hand, it may be one of the best moves you've ever made for the effective delivery of information and training.

Presentation Skills Usage Matrix

Communications Media Type	Examples of Marketing Uses	Examples of Training Uses
Aural	Describing the benefits of purchasing specific goods and/or services	Explaining how to accomplish a specific task
Written	Brochures and pamphlets	Instruction manuals
Visual	Commercials and Demonstrations	Instructional videos for assembly, test, and/or operation of something
Kinesthetic	Test driving an automobile	Dance lessons

Learning Benefits

A 2010 research evaluation performed by the U.S. Department of Education showed that higher education students enrolled in online learning courses performed just as well as those enrolled in on-campus courses and programs.

However, those who combined online learning with occasional on-campus visits actually performed better academically than individuals in traditional, on-campus formats.

How We Really Communicate

- 7% of what we communicate is based on vocabulary
- 38% of what we communicate is based on voice Inflections
- 55% of what we communicate is based on non-verbal behavior

For presentations in a virtual environment, preparation is similar to that of traditional live presentations.

We still need to understand who our audience is and exactly what we want their takeaways to be—that is, what the outcome of the presentation is.

This is no different from traditional presentations. Execution is where the difference lies.

A big way that virtual presentations differ from traditional instructor led trainings (ILT) is in the attention spans of members of your audience.

The different technology packages

The world of Virtual Training and presentations divides up into four generic technology packages: Text, Graphics, Audio and Video.

Here's the breakdown of the choices and elements you'll want to consider when deciding on how you want to use the different systems.

1. Text

- a. E-Mail
    - i. Plain text with and without attachments
- b. HTML
    - i. With embedded links and graphics
    - ii. With attachments
        1. Text
        2. Audio

        3. Video
        4. Files
    c. Video E-Mail
        i. Single Recipient
            1. Live
            2. Pre-recorded
        ii. Multiple recipients
            1. Live
            2. Pre-recorded

2. *Graphics*

- a. PowerPoint
- b. Narrated PowerPoint
- c. Photos and other graphics (Maps, charts, etc.)

3. *Audio*

- a. Teleseminars
    - i. Live
        - 1. Unidirectional broadcast
        - 2. Interactive
    - ii. Recorded
        - 1. Unidirectional broadcast
        - 2. Interactive
    - iii. Audio podcasts
        - 1. Live
        - 2. Recorded
            - a. Linked audio files stored for download and replay

4. *Video*

- a. Webinars
    - i. Live
        - 1. Unidirectional broadcast
        - 2. Interactive
    - ii. Recorded
        - 1. Unidirectional broadcast
        - 2. Interactive
    - iii. Video Podcasts
        - 1. Live
        - 2. Recorded
            - a. Linked audio files stored for download and replay
- b. Video Conferencing
    - i. One-on-one
    - ii. Multiple participants
- c. Video E-Mail
    - i. Single Recipient
        - 1. Live
        - 2. Pre-recorded

ii. Multiple recipients
    1. Live
    2. Pre-recorded

Remember: Attention spans tend to be shorter online than off line.

So, to avoid losing the interest of your audience, you need to create and arrange your content to sustain your viewers' attention.

For instance, consider the slides you will be using. They are the primary visual experience your audience will have, and consequently you need to develop them so they can support your ability to teach a subject or tell a story effectively to your virtual audience.

Moving on, if you think about everything you've seen or heard on radio and television, you'll get a sense of the extreme diversity of programming that you can deliver in a virtual format.

Although virtually any topic can be converted and delivered as a virtual presentation, the most common commercial programming is in the following generic areas:

- Communications
- Business Training Topics
- Safety Training
- General Management
- Accounting
- Computers/Software
- HR Management
- Project management
- Presentation Skills
- Sales/Customer service
- Executive education
- Healthcare
- Time Management
- Organizational development

Regardless of your specific needs, remember that most of these communication models are training sessions with people.

In each case, it's critical to talk with the people online in a natural way. Done wrong, you can actually make friends into enemies so be sure that you're tying the technology with the purpose.

Although it's important what you label the online presentation to get the attention of the target audience, your focus should be on the outcome you want.

The Key Benefits of Virtual Training

For all of the challenges that physical, ILT, event-based training can pose for learners and training decision-makers alike, live virtual training offers a compelling, cost-effective and results-oriented solution to those challenges.

It probably goes without saying, but the benefits of keeping learners in the office or field during training are enormous and bear repeating.

With virtual training, learners are able to:

- Connect training to their jobs and remain engaged with key projects and opportunities
- Better integrate training into their day-to-day activities
- Reduce travel-related hassles and improve quality of work life

Virtual meetings and trainings are real-time interactions that take place over the Internet using integrated audio and video, chat tools, and application sharing.

Imagine that you are an organization that provides training to a number of partners and you have a coalition that stretches across the state.

It can be a real challenge to get everybody in the same room at the same time due to scheduling.

You also know that there are many costs associated with getting everyone together, including gasoline, hotel costs, food, and space to hold the meeting.

The benefits of virtual meetings allow you to reduce costs, more easily find a schedule that works for everyone, and even possibly increase audience size because anyone can attend as long as they have internet access.

The benefit of reaching a larger audience is ideal for an organization that wants to train volunteers across a large geographic area. The online training media might also be a draw for new volunteers that are interested in receiving ongoing training that fits their schedule.

Standardized training for volunteers can be done easily with online training tools because many tools have the ability to record the training to be posted online for access anytime and anywhere.

The benefits of using virtual meetings are extensive. Virtual meeting programs have many features that can serve a variety of needs.

Not all programs offer every feature; however, even the most basic features provide widespread benefits.

1. The first feature of virtual meetings is the ability to share your screen with all meeting attendees.

This feature is the backbone of virtual meetings. If the host pulls up a website, PowerPoint, or Word document, everyone else sees the program on their own screen.

This is extremely beneficial when walking through multiple documents. Other benefits of this feature include developing a collaborative document. Everyone can see whatever the host is writing to ensure their comments were accurately captured.

2. Virtual meetings and trainings allow for multiple presenters in multiple locations.

This feature allows a second trainer to share their screen with all attendees even if they are in a different office or even state.

The process is usually simple and only requires a click of a button to transfer control.

3. The last key feature included in many programs is an interactive element to virtual meetings.

This includes polls, evaluations, or question and answer capabilities.

These features are especially important during lengthy training sessions when it is hard to keep the audience engaged.

An attendee sitting at their desk may be easily distracted by email, phone calls, or people coming to their door.

Polls can also be used to quiz attendees to establish a benchmark of their content knowledge or learning preferences. Evaluations can be done after a session to assess the effectiveness of training.

4. Finally, questions and answers can be used throughout to share ideas or learnings.

To make this whole process a bit simpler to understand and apply, we've divided the virtual world into four categories that we're going to address in the following chapters:

1. Text based communications
2. Graphics applications
3. Audio programming
4. Video programming

The bottom line is that using virtual meetings and trainings will save you time and money.

Part 2 Using PowerPoint Effectively For Training and Collaboration

Why has PowerPoint become so common and what is it?

Since 55% of what we communicate is based on non-verbal behavior, and visual aids are the method of choice for expanding our communications *("A picture is worth one thousand words☺),* it make sense that a tool, specifically designed for enhancing communications through visual content would move to the forefront of training and presentations.

PowerPoint® is a proprietary presentation program developed by Microsoft whose presentations content consists of a number of individual pages or "slides".

Although slides may contain text, graphics, movies, and other objects, which may be arranged freely on the slide, PowerPoint also allows you to use a consistent style in a presentation using a template or "Slide Master".

Effective Use of PowerPoint

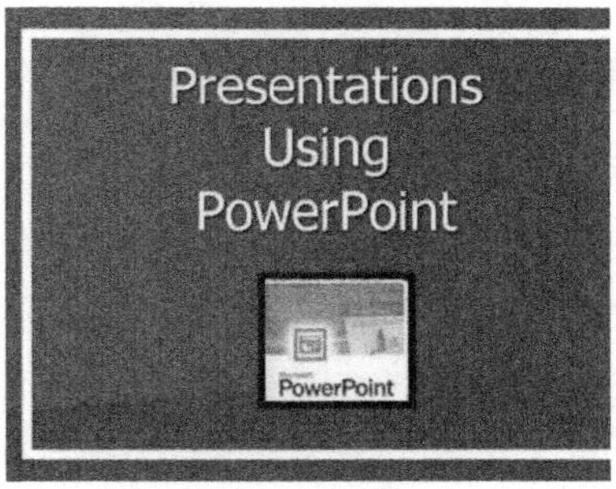

Slide presentation software such as PowerPoint® has become an ingrained part of many instructional settings, particularly in large classes and in courses more geared toward information exchange than skill development.

These, of course, include virtual presentations and on-line learning programs.

PowerPoint can be a highly effective tool to aid learning, but if not used carefully, may instead disengage students and actually hinder learning.

Potential benefits of using presentation graphics include:

- Engaging multiple learning styles
- Increasing visual impact
- Improving audience focus
- Providing annotations and highlights

Although there are many potential benefits to PowerPoint, there are several issues that could create problems or disengagement:

- PowerPoint was designed to promote simple persuasive arguments so design for critical engagement, not just for exposure to a "point."
- Presentation graphics should be about learning, not about presentation.
- PowerPoint presentations should help your audience organize their thinking and notes, not just "be" the notes.

This is a particular danger with members of your audience who grew up accustomed to receiving PowerPoint notes to study from. Your audience may require convincing that notes should be taken beyond what is already on the slides.

"Death by PowerPoint"

Death by PowerPoint is a phenomenon caused by the poor use of presentation software.

Key contributors to death by PowerPoint include confusing graphics, slides with too much text and presenters whose idea of a good presentation is to read 40 slides out loud.

Death by PowerPoint is easily recognized by observing the audience members' glazed eyes, furtive use of smartphones and trips to the bathroom.

In virtual presentations, this is hard to observe but you'll have to assume that if the presentation drones on forever, the audience has zoned out and is no longer paying any attention to your material.

When an audience remains emotionally disconnected from the content that is being presented, there is a good chance that the presenter has either not spent enough time and effort thinking about which key points he wants the audience to take away -- or he has spent entirely too much time and effort setting up the presentation in PowerPoint, incorporating every feature and display option the software provides.

Death by PowerPoint can be avoided if the presenter uses the technology as a visual aid to enhance what is being said, instead of relying on the technology to serve as the focus of the presentation.

Less is STILL more!

- How long is your attention span?
- Have we lost you already?
- Are you motivated to pay attention to our message?
- What is in this for you?

These are questions that we must face when presenting information to any audience.

We must focus here on our target audience who lives in a world of media and information overload.

Movies, news programs, and the inundation of electronic information have forced us all – our audience included – to develop ever shorter attention spans as a means of survival.

How can we as presenters prepare presentations that capture and hold audience attention when providing them with information critical to our case?

Our goal is to provide you with helpful and practical suggestions that can be used in any case, large or small, to make visual presentations more interesting and capable of holding a short attention span.

So set your focus here – I said here, please.

Some Facts

Have you watched an old movie lately? The pace seems painfully slow. We have adjusted our sense of timing with the change of the "average shot length" (ASL).

Film scholar Barry Salt has studied this metric and published way more information that you might want to know on his web site, www.cinemetrics.lv.

Here's the conclusion based on his observation and research – average shot length has shortened.

- The action film Ben Hur, released in 1959, has an ASL of 8 seconds overall, with the heart pounding rowing scene shortened to 1.2 seconds.

- The Bourne Supremacy, released in 2004, has an ASL of 2.4 seconds on average.

The difference in pace is stunning

Our audiences are instant gratification junkies who get the news in capsule format and who stand in front of their microwaves, look at their watches, and yell *"hurry up!"*

When you deliver a virtual presentation you're dealing with the same mindset and overall impatience so you MUST consider this when creating your programs.

It must move rapidly, be informative, and entertaining to the audience or they'll tune out and miss the entire reason for the program.

The key to success is to seriously consider the question: Can VS. Should

When creating your virtual presentation, you'll find that are a lot of things you can do but the question is: Should you?

If it really pertains to the topic and more importantly, is it important to the audience, by all means, put it into the presentation.

If it's irrelevant or something that is no interest to the audience, then leave it out.

Creating and delivering presentations using PowerPoint

I started this section of the program by referencing PowerPoint™ because if you were to survey 100 people, 101 of them would equate virtual training and content delivery with PowerPoint™ and although it's not the only tool in set, it's undoubtedly the most widely known, used and unfortunately, abused.

Let's start by saying that many presentations use visual aids and showing things to an audience during a presentation is as old as public speaking.

In nearly all cases, showing an audience a physical thing, an actual object, is the best way to engage an audience's attention plus if you can show a picture, diagram, or sketch, you eliminate the need for the recipient to try and imagine what you're describing thereby eliminating a misconception or misinterpretation of what it is that you're describing..

But when this isn't possible, presentation software like PowerPoint allows the modern public speaker to show things to an audience on a large screen.

What has been turned upside-down over the past decade's spread of PowerPoint™, for most PowerPoint users, is that the "speech" is now mostly what's on the screen, rather than what is spoken.

In other words, the proper relation of the illustration tool to the speech has been reversed.

In the opinion of many people, this has tragically damaged the art of public speaking. No one can imagine Abraham Lincoln nor Martin Luther King, Jr., needing PowerPoint™.

But today many people who give oral presentations cannot imagine doing so without PowerPoint™.

Before you grab a rope and decide to lynch me for PowerPoint™ treason, pay close attention to the following information and temper your use of the software judiciously.

When used as a visual aid, PowerPoint™ can be one of the most useful tools in your arsenal of delivery devices. On the other hand, you can also use it as an instrument of self-destruction in your quest to deliver solid, usable content.

In the interest of restoring some balance to the use of PowerPoint™, without rejecting its use altogether, here are some suggestions for how to use PowerPoint effectively.

This is one of those areas where two people will have a dozen opinions so rather than present these as absolutes; I'm going to refer you to my key word, appropriate.

By all means, look through and review all of these tips, tricks, and suggestions and pick and choose that are appropriate for your presentation because realistically, you can't follow all of them and remain sane because several of them are at odds with each other.

Ten Thoughts about How to Use PowerPoint Effectively

1. **PowerPoint, when displayed via a projector, is a useful tool for showing audiences things that enhance what the speaker is saying**. It is a useful tool for illustrating the content of a speech, such as by showing photos, graphs, charts, maps, etc., or by highlighting certain text from a speech, such as quotations or major ideas. It should not be used as a slide-show outline of what the speaker is telling the audience.

2. **Slides used in a presentation should be spare, in terms of how much information is on each slide, as well as how many slides are used.** A rule of thumb is to put no more than eight lines of text on a slide, and with no more than eight to ten words per line. In most cases, less is more, so four lines of text is probably better. Don't display charts or graphs with a lot of information—if it's useful for the audience to see such things, pass them out as handouts.

3. Unless you're an experienced designer, don't use the transition and animation "tricks" that are built into PowerPoint, such as bouncing or flying text. By now, most people roll their eyes when they see these things, and these tricks add nothing of value to a presentation.

4. Above all, use high-contrast color schemes so that whatever is on your slides is readable. Unless you are a talented graphic designer, use the templates that come with PowerPoint or Keynote, and keep it simple—high concept design in a slide presentation doesn't help in most circumstances, unless you're in the fashion or design fields. If you use graphics or photos, try to use the highest quality you can find or afford—clip art and low-resolution graphics blown up on a screen usually detract from a presentation.

5. Rehearse your PowerPoint presentation and not just once. Don't let PowerPoint get in the way of your oral presentation, and make sure you know how it works, what sequence the slides are in, how to get through it using someone else's computer, etc. Make sure that you can deliver your presentation if PowerPoint is completely unavailable; in other words, make sure you can give your speech without your PowerPoint presentation.

6. Get used to using black slides. There are few speeches that need something displayed on the screen all the time. If you include a black slide in your presentation, your audience will refocus on you, rather than on the screen, and you can direct them back to the screen when you have something else to show them. Put a black screen at the end of your presentation, so that when you're done, the PowerPoint presentation is finished and off the screen.

7. Concentrate on keeping the audience focused on you, not on the screen. You can do this by using slides sparingly, standing in front of the audience in a way that makes them look at you, and, if possible, going to the screen and using your hand or arm to point out things on a slide. If you expect to be using PowerPoint a lot, invest in a remote "clicker" that lets you get away from the computer and still drive your presentation. If you don't have one of those, it's better to ask someone to run the presentation than to be behind a screen and keyboard while you talk.

8. If you show something on a computer that requires moving the cursor around, or flipping from one screen to another, or some other technique that requires interaction with the computer itself, remember that people in the audience will see things very differently on the projection screen than you see them on the computer screen. Keep motion on the screen to a minimum, unless you're showing a movie or a video. It's better to show a static screenshot of a Web page, embedded on a slide, than to call up the Web page in a browser on a computer. If you want to point out something on a Web page, go to the screen and point at it—don't jiggle the cursor around what you want people to look at: their heads will look like bobble-headed dolls.

9. Don't "cue" the audience that listening to your speech means getting through your PowerPoint presentation. If the audience sees that your PowerPoint presentation is the structure of your speech, they'll start wondering how many slides are left. Slides should be used asynchronously within your speech, and only to highlight or illustrate things. Audiences are bored with oral presentations that go from one slide to the next until the end. Engage the audience, and use slides only when they are useful.

10. Learn how to give a good speech without PowerPoint. This takes practice, which means giving speeches without PowerPoint. Believe it or not, public speaking existed before PowerPoint, and many people remember it as being a lot better than it is now. A few people use presentation software in extremely effective ways—Steve Jobs and Stanford Law Professor Lawrence Lessig are two examples. Al Gore's use of Keynote in the movie "An Inconvenient Truth" was a good model. But these three examples don't look at all like the way most people use PowerPoint. Avoiding bad PowerPoint habits means, first and foremost, becoming a good public speaker.

Tips for creating better PowerPoint™ presentations

Images

Images are key elements of every presentation.

Your audience has ears and eyes – they'll want to see what you're talking about, and a good visual cue will help them to understand your message much better.

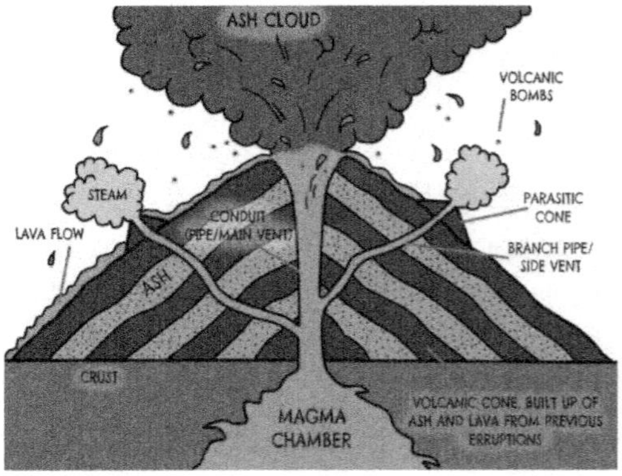

Add Images

- Have more images in your slides than text.
- But do not use images to decorate!
- Images can reinforce or complement your message.
- Use images to visualize and explain.
- A picture can say more than a thousand words.
- If you don't have your own images, you can use royalty free art sites like clipart.com and keep copyrights in mind.

Animations & Media

In animations, there is a fine line between a comic or professional impression.

However, animations can be rather powerful tools to visualize and explain complicated matters.

A good animation can not only improve understanding, but can also make the message stick with your audience.

- Don't Be Silly
- Use animations and media sparingly.
- Use animations to draw attention, for example to your Take Home Message.
- Use animations to clarify a model or emphasize an effect.

Target & Content

Your target i.e. your audience, defines the content of your presentation.

For example, you won't be able to teach school kids about the complicated matters of economy.

However, you may be able to explain what economy is in the first place and why it is important.

- Keep Your Audience In Mind
- What do they know?
- What do you need to tell them?
- What do they expect?
- What will be interesting to them?
- What can you teach them?
- What will keep them focused?

Answer these questions and boil your slides down to the very essentials.

In your talk, describe the essentials colorfully and choose your weapons i.e. text, images and animations wisely (see above).

If you lose the audience's attention — it won't matter how ingenious your design or how brilliantly you picked colors and keywords.

Practice

A well-prepared and enthusiastic talk will help you convince your audience and maintain their attention. There are some key points that define a good talk.

- Know your slides inside out.
- Speak freely.
- Speak with confidence – loud and clear.
- Don't speak too fast.
- Maintain eye contact with the audience.

Structuring Your Talk:

- Preparing a talk always takes far longer than you anticipate. Start early!
- Write a clear statement of the problem and its importance.
- Research. Collect material which may relate to the topic.
- Tell a story in a logical sequence.
- Stick to the key concepts. Avoid description of specifics and unnecessary details.

If you are making a series of points, organize them from the most to the least important. The less important points can be skipped if you run short of time.

Keep your sentences short, about 10-20 words each is ideal. This is the way people usually talk.

Strive for clarity.

- Are these the best words for making your point?
- Are they unambiguous?
- Are you using unfamiliar jargon or acronyms?

Preparing Your Slides:

- Presentation Design
- Let the picture or graphics tell the story - minimize the use of text.
- Don't overload your slides with too much text or data.
- FOCUS. In general, using a few powerful slides is the aim.
- Type key words in the PowerPoint Notes area listing what to say when displaying the slide. The notes are printable.
- Number your slides and give them a title.
- Prepare an Agenda or Table of Contents slide. You can reuse the same slide at the end of the presentation by changing the title to Summary.
- Prepare a company logo slide for your presentation.
- You can add a logo and other graphics to every slide using the slide master feature or by adding them to the footer.
- Proofread everything, including visuals and numbers.
- Keep "like" topics together.
- Strive for similar line lengths for text.

Visual elements

- A font size of 28 to 34 with a bold font is recommended for subtitles. The title default size is 44. Use a san serif font for titles.
- Use clear, simple visuals. Don't confuse the audience.
- Use contrast: light on dark or dark on light.
- Graphics should make a key concept clearer.
- Place your graphics in a similar location within each screen.
- To temporarily clear the screen press W or B during the presentation. Press any key to resume the presentation.

Text

- Font size must be large enough to be easily read. Size 28 to 34 with a bold font is recommended.
- It is distracting if you use too wide a variety of fonts.
- Overuse of text is a common mistake.
- Too much text makes the slide unreadable. You may just as well show a blank slide. Stick to a few key words.
- If your audience is reading the slides they are not paying attention to you. If possible, make your point with graphics instead of text.
- You can use Word Art, or a clip art image of a sign, to convey text in a more interesting way.

Numbers

- Numbers are usually confusing to the audience. Use as few as possible and allow extra time for the audience to do the math.
- Numbers should never be ultra-precise: "Anticipated Revenues of $660,101.83" looks silly. Are your numbers that accurate? Just say $660 thousand. "The Break Even Point is 1048.17 units. Are you selling fractions of a unit?
- Don't show pennies. Cost per unit is about the only time you would need to show pennies.
- If you have more than 12-15 numbers on a slide, that's probably too many.
- Using only one number per sentence helps the audience absorb the data.

Statistics

- Use the same scale for numbers on a slide. Don't compare thousands to millions.
- When using sales data, stick to a single market in the presentation. Worldwide sales, domestic sales, industry sales, company sales, divisional sales, or sales to a specific market segment are all different scales. They should not be mixed.
- Cite your source on the same slide as the statistic, using a smaller size font.

Charts

- Charts need to be clearly labeled. You can make more interesting charts by adding elements from the drawing toolbar.

- Numbers in tables are both hard to see and to understand. There is usually a better way to present your numerical data than with columns and rows of numbers. Get creative!
- PowerPoint deletes portions of charts and worksheets that are imported from
- Excel, keeping only the leftmost 5.5 inches. Plan ahead.

Backgrounds

- Backgrounds should never distract from the presentation.
- Add a design style or a color to the background.
- Backgrounds that are light colored with dark text, or vice versa, look good. A dark background with white font reduces glare.
- Colors appear lighter when projected. Pale colors often appear as white.
- Consistent backgrounds add to a professional appearance.
- For a long presentation, you may want to change background designs when shifting to a new topic.

NOTE: See the section below on template vs custom backgrounds

Excitement

- Slides for business presentations should be dull! You don't want to distract the audience.
- Sounds and transition effects can be annoying. Use sparingly.
- Animation effects can be interesting when used in moderation.
- Too much animation is distracting.
- Consider using animated clip art
- Consider using custom animation
- You can insert video and audio clips into PowerPoint.
- You can also insert hyperlinks.

Hints for Efficient Practice:

- Timing - Practicing Your Presentation,
- Talk through your presentation to see how much time you use for each slide.
- Set the automatic slide transition to the amount of time you want to spend discussing each slide.
- Are you using the right amount of time per slide? Decide which slides or comments need alteration to make your presentation smoother.
- Change the automatic slide transition settings for individual slides to fit the amount of time needed for that slide and practice again. Are you still within the time limit?
- Decide if you want to remove the automatic slide transition feature before giving the presentation.

Content

- Make a list of key words/concepts for each slide
- Read through the list before you begin.
- Don't attempt to memorize your text;
- Your words will probably be different each time you practice.
- Think about the ideas, and your words will follow naturally.

Delivering Your Talk:

- Pre-Talk Preparation
- Plan to get there a few minutes early to set up and test the equipment.
- Dress appropriately for your audience.
- Turn off your cell phone.

Handouts:

- Edward Tufte, the leading expert on visual presentation techniques, advises speakers to always prepare a handout when giving a PowerPoint presentation.
- Make about 10% more handouts than you expect to use.
- Distribute handouts at the beginning of your talk.

Opening:

- Jump right in and get to the point.
- Give your rehearsed opening statement; don't improvise at the last moment.
- Use the opening to catch the interest and attention of the audience.
- Briefly state the problem or topic you will be discussing.
- Briefly summarize your main theme for an idea or solution.

Speaking

- Talk at a natural, moderate rate of speech
- Project your voice.
- Speak clearly and distinctly.
- Repeat critical information.
- Pause briefly to give your audience time to dig new slide.
- Don't read the slides aloud. Your audience ca you can talk.
- If you plan to write on the slides to empha presentation, practice ahead of time. To selec during the presentation.

Body Language

- Keep your eyes on the audience
- Use natural gestures.
- Don't turn your back to the audience.
- Don't hide behind the lectern.
- Avoid looking at your notes. Only use them as reference points to keep you on track. Talk, don't read.

Questions

- Always leave time for a few questions at the end of the talk.
- If you allow questions during the talk, the presentation time will be about 25% more than the practice time.
- You can jump directly to a slide by typing its number or by right-clicking during the presentation and choosing from the slide titles.
- Relax. If you've done the research you can easily answer most questions.
- Some questions are too specific or personal. Politely refuse to answer.
- If you can't answer a question, say so. Don't apologize. "I don't have that information. I'll try to find out for you."

Length:

- To end on time, you must PRACTICE!

- When practicing, try to end early. You need to allow time for audience interruptions and questions.

Demeanor:

- Show some enthusiasm. Nobody wants to listen to a dull presentation. On the other hand, don't overdo it. Nobody talks and gestures like a maniac in real life. How would you explain your ideas to a friend?
- Involve your audience. Ask questions, make eye contact, and use humor.
- Don't get distracted by audience noises or movements.
- You'll forget a minor point or two. Everybody does.
- If you temporarily lose your train of thought you can gain time to recover by asking if the audience has any questions.

Conclusion:

- Concisely summarize your key concepts and the main ideas of your presentation.
- Resist the temptation to add a few last impromptu words.
- End your talk with the summary statement or question you have prepared.
- What do you want them to do? What do you want them to remember?

Consider alternatives to "Questions?" for your closing slide.

A summary of your key points, a cartoon, a team logo, or a company logo may be stronger.

Remote controls

PowerPoint presentations should be used in conjunction with a remote control so you don't have to keep the computer's keyboard in front of you.

They are made up of two parts: transmitter and USB receiver

Most of them combine the function of a laser pointer with slide forward, slide back, and the better ones blank the screen.

They generally operate up to 75 feet and they're easy to use.

Look in the resources section for the models and links to the ones we recommend.

Templates Vs Custom Backgrounds

One way of creating a presentation is to start from scratch. However, you can save time and effort if you base your new presentation on a template.

A template defines the background, font styles, colors and sizes for your placeholders, as well as selected bullets that match the template.

Using a template means that you don't need to worry about defining those aspects yourself

PowerPoint comes with a number of stylish templates built in that you can use immediately, and you can download additional ones from office.com or from the other resources in this manual.

Microsoft PowerPoint comes with resources that include a typeface library, themes and templates. These elements work together to provide cohesive sets of ready-to-customize slide masters that you can use to speed presentation development when you need to meet an urgent deadline for a customer or staff meeting.

In some circumstances, however, you may prefer to step out from behind the preset choices and create a PowerPoint template that's unique to your company or business. But starting from scratch in PowerPoint does present its own set of pluses and minuses.

Professionally designed PowerPoint templates feature themes that rely on harmonious palettes of color, typefaces chosen for easy readability and a full range of slide types to suit virtually any presentation need. These options speed the process of setting up individual slides and crafting a visually cohesive slide show.

If you choose a preset template and wind up spending a great deal of time customizing it to meet your specific needs, however, its value drops accordingly.

Company Style - Preparing PowerPoint presentations that feature your company style may require you to set up a custom template based on your corporate colors, signature typefaces, in-house style rules and other governing principles, which prefabricated templates simply can't match.

Of course, you could commission the creation of a template rather than design it yourself, but you may prefer to avoid the extra expense.

If your company standards dictate that all your materials feature your logo, look and feel, however, only custom development enables you to avoid violating your own in-house rules.

Originality - The templates that ship with Microsoft PowerPoint, and those you can download free from the Microsoft website have become more than a little overexposed, especially PowerPoint's default template.

If you're tired of crafting predictable-looking presentations and want something with greater story-telling value than the standard bulleted slides, you may need to start from scratch to achieve the look you want.

The task of pulling together a truly original template, however, could be daunting if you lack design training or aren't sure of the look you want.

Why using templates can be a bad idea

For as great an idea as templates can be, they can also become boring and cause the audience to ignore the material being presented.

The thematic appearance of a template presentation may be great if you're trying to stay in sync with a corporate or organizational image but if you're really using PowerPoint as a visual presentation system, you'll recognize that it's designed to allow you to present visual information and NOT to be a cloned series of mind-numbing slides.

In short, look closely at what you're trying to convey to your audience and structure visuals accordingly.

Additional Recommended PowerPoint® Tips and Techniques

For as popular as PowerPoint has become, there are hundreds, perhaps thousands, of tips and ideas about how to use it more effectively.

Since we are primarily concerned with delivering a good presentation and not impressing people with our computer skills, we've boiled the tips down to the following items for you to consider

- Start with the end in mind.
- Stick with what works. Use a template.
- Use the Outline view to write content quickly.
- Use images to liven things up.
- Present your data in a chart.
- Create your own template for repeated presentations.
- Color-coordinate your materials.
- Instead of handouts, make your presentation available on the web.
- Practice, practice, practice. Use Rehearse Timings.

Working in Outline view

If you're not a skilled user of PowerPoint, you'll definitely want to use their Outline View to help you create your program.

- The Outline view displays your presentation as an outline made up of the titles and main text from each slide.
- Each title appears on the left side of the pane that contains the Outline tab, along with a slide icon and slide number.
- The main text is indented under the slide title.
- Graphic objects appear only as small notations on the slide icon in Outline view.

Working in Outline view is particularly handy if you want to make global edits, get an overview of your presentation, change the sequence of bullets or slides, or apply formatting changes.

When you save your presentation as a Web page, the text on the Outline tab becomes a table of contents so that you can navigate through the slides.

Free and low cost PowerPoint Support Sites and Materials

Here are a variety of sites with some great templates and add-ons that will assist you in creating better PowerPoint Presentations.

Free Open Office

If you're not interested in purchasing or maintaining the latest version of PowerPoint, seriously consider downloading and installing open office which contains open source equivalents of the Microsoft Office Suite including their alternative called ***Impress™.***

It includes multimedia presentation tools, such as special effects, drawing capability, sound and video clips along with animation.

http://downloadser.com/openoffice

Impress, is capable enough, but it's not going to convert any naysayers.
The main issue with ***Impress™*** is that it still struggles occasionally when importing presentations made in PowerPoint like some missing fonts and other rendering problems.

That said, if you're creating presentations from scratch, Impress does it pretty well, albeit without quite as much flair as PowerPoint.

That pretty much equates to a lack of certain slide animations and no support for video exports.

Impress also doesn't have animated diagrams and lacks the ability to collaborate on presentations as a group.

It's not all bad news for Impress, though. Impress can export presentations in a ton of different formats, which makes sharing presentations a heck of a lot easier because you don't need to rely on someone else having PowerPoint.

Regardless, both feel pretty similar to use, and while neither will magically give you an Academy Award-winning presentation; they're both equally easy to start.

The Bottom Line

Open Office's Impress is a perfectly capable program, but if you're working with people who usually use PowerPoint you might run into some compatibility issues still.

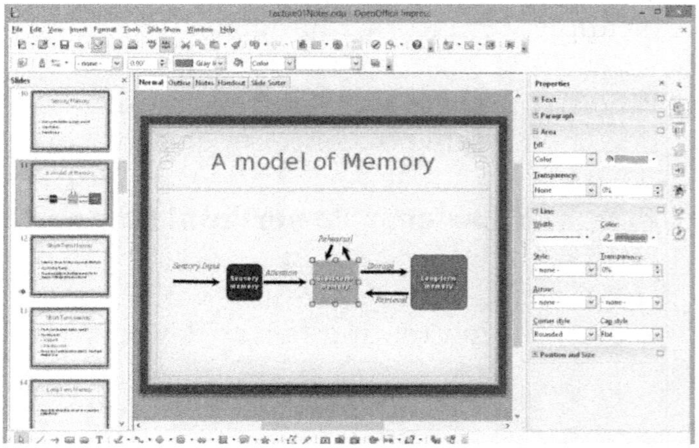

If you're just making presentations for your own purposes, then Impress should work for you.

OpenOffice is easy to learn, and if you're already using another office software package, you'll take to OpenOffice straight away. And if you already have files from another office package - OpenOffice will probably read them with no difficulty.

Best of all, Apache OpenOffice can be downloaded and used entirely free of any license fees. You may install it on as many computers as you like. You may make copies and give them away to family, friends, students, and employees - anyone you like.

Desktop Timer

This is a simple utility that allows you to create multiple timers on your desktop that you might want to use if you're timing your presentation or if you're delivering timed exercise with an audience.

Once installed, Free Desktop Timer appears as an icon in the system tray, ensuring that it's always handy when you need it like timing an exercise or limiting the time someone had to present material during your presentation.

http://download.cnet.com/Free-Desktop-Timer/3000-2350_4-75415517.html#ixzz32Nco5LN3

Free Templates for PowerPoint

Although PowerPoint comes with a bunch of templates, there are lots of other sites that provide a boatload of free templates that you might find more useful in your particular solution.

Look through their catalog of offering and pick whatever you think will best serve your purposes.

Remember that since they are templates that they can be modified, customized and adapted for you own specific needs or organizational branding.

www.free-powerpoint-templates-design.com/free-powerpoint-templates-design/

This is one of the most useful PowerPoint template website providing a huge collection of Free PowerPoint templates that you can download for free and use for your personal Presentations. The templates are appropriate for business and lecture room presentations on education, health, and training.

www.presentationmagazine.com/free_powerpoint_template.htm

This site has collection of 59,424 free PowerPoint templates, themes and PowerPoint Backgrounds.

You can browse all of their PowerPoint templates or select them by category or color or by tag. They also have a range of PowerPoint Backgrounds.

www.presentationmagazine.com

This is the site for Presentation Magazine and if you're serious about doing better job of creating and delivering presentation, you'll want to bookmark this site and check out what tips, advice, and tools they offer on a regular basis.

www.presentationmagazine.com/editable-book-cover-14691.htm

This is a sub-site for Presentation Magazine that deals with creating custom book covers for posting on the web.

If you're tired of seeing your on-line reports with a plain black and white title page, you MUST check this site out.

www.presentermedia.com/?gclid=CJek89HFtL4CFSgQ7AodJgUAWA

Tis site has an insane number of free templates that feature animation, video, and complete template sets that are designed as attention getters and for standalone presentations rather than as support for your live presentation.

www.free-power-point-templates.com

This is Free PowerPoint Templates, one of the most visited PowerPoint templates resource websites containing a free collection of over 1,000 PowerPoint design templates, PowerPoint pictures, animations and illustrations

Part 3 Using Audio for Training and Collaboration

As training goes high tech and virtual in delivering content, presenters are looking for ways to present their traditional material online and audio is the place to start.

Background of the medium

Let's start with the audiobook and although most people will call anything spoken and recorded an audiobook, technically, it's a recording of a text being read.

A reading of the complete text is noted as "unabridged," while readings of a reduced version or abridgement of the text are labeled as "abridged."

Spoken audio has been available in schools and public libraries and to a lesser extent in music stores since the 1930s and many spoken word albums were made prior to the age of videocassettes, DVDs, compact discs, and downloadable audio.

During the 1930s, many people wanted to broadcast and record themselves and in the 1940s, there were actually recording booths where you could record a letter, sing a song, or just speak and out would pop a record.

It was not until the 1980s that the medium began to attract book retailers, and then book retailers started displaying audiobooks on bookshelves rather than in separate displays.

The term "talking book" came into being in the 1930s with government programs designed for blind readers, while the term "audiobook" came into use during the 1970s when audiocassettes began to replace records.

In 1994, the Audio Publishers Association established the term "audiobook" as the industry standard.

With the development of portable cassette recorders, audiotapes had become very popular and by the late 1960s libraries became a source of free audiobooks, primarily on vinyl records but also on cassettes.

Instructional and educational recordings came first, followed by self-help tapes and then by literature.

In 1975, Olympic gold medalist, Duvall Hecht founded *Books on Tape, Inc.* as a direct to consumer mail order rental service for unabridged audiobooks and expanded their services selling their products to libraries and audiobooks gaining popularity with commuters and travelers.

By the middle of 1980s the audio publishing business grew to several billion dollars a year in retail value.

With the advent of the Internet, broadband technologies, new compressed audio formats and portable media players, the popularity of audiobooks has increased significantly.

As noted above, audiocassettes and now CDs have made this a commercially viable marketplace because of its portability.

When the MP3 player, in the form of the iPod became commercially available, the market has literally exploded and now if you're in a position to create and offer audio products, there is a virtually unlimited marketplace for whatever you can supply

Why audio in a video centered world?

You might think I'm crazy to suggest that you make audio products instead of creating video.

It's a fair point especially when you consider that video is everywhere right now and it's a real force to be reckoned with.

Its popularity – particularly with Internet Marketers – is huge and pretty much anyone who's anyone is using some form of video in their marketing mix with impressive results.

Nevertheless, Audio Is Awesome!

In fact with so many people concentrating on creating video, it gives you the perfect opportunity to stand out from the "me too!" crowd by creating your own great-sounding audio products.

It's worth remembering that just because everyone else seems to be using video it doesn't mean you have to blindly follow the herd.

In fact, there are many advantages to creating audio products instead of video – many of which you might not have even thought about.

Audio Products Are Easier To Make

Firstly, creating your own audio products has, and probably always will be, quicker and easier than creating video.

With audio you don't have to worry about getting the lighting right, or spending hours learning complicated video editing software or finding out how to upload and embed it on your website.

Another of the major downsides to video is having to with your appearance on camera. I don't mind appearing on camera as I've been doing it for most of my adult life but it's not something that most people are entirely comfortable with or enjoy doing.

But with audio products it's so much easier – all you need to do is speak... and we all do that every day whether it's on the phone or face-to-face, so it's not like you have to acquire a new skill or anything like that.

Hit *Record* and Have A Product Within An Hour

With audio products you simply hit the 'record' button and start speaking.

Once you've run out of things to say you just hit 'stop' and your product is pretty much ready.

I know that's a very simplistic approach, but it's true.

Don't believe me? Well think about this for a second...

If you plan to record yourself speaking for 20 minutes on your chosen topic/niche, or perhaps do an interview with a guest of a similar duration, it means that your audio product will be ready in little more time than it takes you to actually record it.

You will, of course, need to do a few minutes of preparation before the recording starts to allow you to write some questions but that should take no more than 30-40 minutes if you really focus (remember to switch off your phone and log-out of your emails).

You'll also need a few minutes afterwards to edit the audio (if it needs it) and to convert it into MP3 format for uploading to your website or for burning to CD.

But that's it! Literally everything you need to create your first audio product in round about an hour.

Where do you get audio recordings?

There are a variety of ways to get your material into an audio format including

Recordings of live performances –

Live performances have been recorded and offered to the public since the days of vinyl records. In the 1960s, many of the greatest speakers such as Earl Nightingale, Dr. Norman Vincent Peal, and a host of others began to record their live speeches and presentations and offer them to the public.

Since then, tens of thousands of audio presentations from motivational speeches to sales seminars, and standup comedy routines to instruction manuals have been recorded and sold.

If you are skilled platform presenter and speak to live audiences, you can record your presentations, edit them, commit them to a distribution medium, and offer them to the public.

Recordings of printed materials - This is the classic definition of an audio book where the narrator actually reads from a printed book.

Depending on the type of book and the subject matter, the narrator may be called upon to add different voices, accents, and a range of emotions in order to make the audio book come alive.

This is a commonplace activity with a full range of books being recorded for those who are visually impaired

More importantly, because audio is the least invasive of all mediums, that is, you can multitask while listening; it lends itself perfectly to a full range of informational, motivational, inspirational books for those who don't have the time to dedicate to actually reading the material or for those who find it difficult because of their circumstances.

For example, Michelle Pardes is a lymphoma cancer survivor who wrote a book about her experiences as a way to help others who are undergoing cancer treatment.

Although the book is spectacular in its content and highly motivational for those who are undergoing chemotherapy and other forms of invasive treatment, it's very difficult for someone who has nausea and other side effects to read a book.

Michelle recorded her entire book and makes it available on CD so that those who need the material and don't feel well enough to read.

Scripted material - some audio material is created specifically to be delivered in an audio format.

One of the most successful examples of this technique is known as the *expert question-and-answer session.*

An expert on a given topic writes down most the interesting and commonly asked questions about their area of expertise.

Using a professional speaker as an interviewer to ask the questions, the expert answers the questions, the whole session is recorded and edited, and then committed to a reproduction medium, and offered to the buying public.

Preparing Quality Audio Segments

For as simple and commonplace and audio appears to be, if people can't understand what is being said on an audio segment, then there is no point in using it.

The quality of an audio segment is magnified – for better or for worse – when it is amplified through a sound system.

A good quality audio segment sounds clear and crisp, and a poor quality audio segment sounds even worse and more distorted when it is amplified.

First, one of the biggest factors that determine the quality of the audio segment is the sample rate of the recording.

- The sample rate refers to the number of times that the audio signal is sampled as it is recorded into digital form.
- The higher the sampling rate, the better the quality of the sound will be.
- CD quality sound has a sampling rate of 44,100 while the telephone uses a sample rate of 8000 (measured in Hertz).
- You want to get as high a sample rate as possible on the original recording because increasing sample rate through conversion is very difficult to do.

Second, another factor in quality is whether the recording is in stereo (both channels recorded) or mono (only one channel recorded).

- A stereo recording has better sound because it is closer to sounds in the real world where we hear in both ears.

Third, a higher quality sound recording will create a larger file on your recording device and a larger audio file on your computer, so this may be a trade-off you need to keep in mind if you are distributing the file on a format such as diskette or CD.

The fourth component is the microphone itself.

Although you can probably get away with the built in microphone on your laptop computer, I wouldn't recommend it unless it's an emergency.

A decent quality microphone for voice recording can be had for less than $20 bucks and it's a worthy investment if you want to make quality recordings.

The best audio recorder you can get to record your audio is your computer because it's records in full digital media with no losses or distortion.

Just about any Macintosh or Windows computer will work as long as it has an audio card installed or built in and coupled with Audacity™, you'll have studio quality immediately.

Regardless of the computer you use, the microphone is where the sound originates and if you want good quality sound, you'll need to get a good microphone.

You can get a simple headset microphone or plastic stand mike for less than $5 from discount locations and they'll work OK when you record your voice in a quiet location.

If you're serious about sound quality, there's a great microphone that's been configured specifically for creating digital voice recordings called the SF-90 Microphone.

You'll find it at www.tmart.com/SF-920-Microphone-for-PC-Laptop-Notebook_p230722.html for about $20.

I personally use and recommend the Blue Microphone Yeti USB Microphone for about $99 plus a good set of headphones to monitor your work http://tinyurl.com/3wj86qy.

I've also had outstanding results with the Microsoft LifeChat LX-3000 Headset and boom microphone.

Since its hands free, you can move your head anywhere you want to make it comfortable and easy for you to record. It's about $25 bucks and is available from Amazon.com and dozens of other outlets.

If you prefer to record your material using a dedicated miniature digital recorder, I recommend the Sony Digital Voice Recorder, Model: ICDPX333.

It's about $60 bucks at places like Best Buy™ and is compatible both the Windows® and Macintosh® operating systems so you can import the audio and edit it with Audacity™.

If you're going to be doing a lot of professional recording in a variety of locations, you'll want to investigate wireless microphones and mixers.

Although you'll find a lot of them on-line at reasonable prices at places like Amazon.com, I strongly recommend that you visit a store like the Guitar Center™ and actually test a few of the systems to see which ones will best suit your needs.

Another critical factor that many people forget about is the echo that occurs in a room when you record.

The farther away from the microphone you are, the more echo and ambient noise you introduce into the recording and it's for this reason that you see sound crews holding a mike above a celebrities head and on-scene reporters holding a handheld mike close to their mouth.

For maximum sound quality when recording, narrating, or overlaying sound at the computer, always use a desktop microphone that you can position close to your mouth to eliminate as much echo and ambient background noise as possible.

This also holds true for any audio recording you decide to do including audio and video podcasts and anything that involves recording your voice or the voices of others.

When it comes to video recording, you MUST have a camcorder that accepts a separate audio input. Otherwise, you're going to have noise, echo, and a badly recorded audio track. (More about the cameras later on)

You can look on EBay and find wireless microphone systems for $10 bucks and up and they might work OK *(Definitely a lot better than using the built in microphone on the camera)* but don't count on it.

I've tried a bunch of the really cheap units and you get what you pay for – a lot of poor quality sound, static, and interference from everything electrical in the area.

You need to spend a few dollars on quality wireless mike system that has good quality and isn't affected by outside electrical noises.

You can go crazy on the high end of things (Like with a Sennheiser Studio Quality system) but for most people, Nady, Samson, and Azden microphones will fit the bill nicely.

I suggest you start with the Azden WMS-PRO Wireless Microphone System which has a wireless belt transmitter and both the lavaliere and handheld microphones plus the wireless receiver.

The microphones will work reliably up to 50 feet or more from the receiver and you can find them on EBay for around $100.

You can use the Azden for straight audio recording into a computer as well because it's a quality system and offers you the opportunity to sit, stand, or walk around as you're recording.

On the other hand, if you're tied to a desk, and especially if you are conducting interviews or recording a group of people, we recommend that you invest is a good microphone that will cover multiple environments.

There are a couple of solutions and the one we recommend as previously mentioned is the Yeti Microphone.

You can find it on EBay for less than $100 bucks and it has some spectacular features that make it the ideal universal microphone for most applications:

The Yeti features an innovative triple capsule array, allowing for recording in stereo or your choice of three unique patterns, including cardioid, omnidirectional, and bidirectional, giving you recording capabilities usually requiring multiple microphones.

The Yeti utilizes a high quality analog-to-digital converter to send incredible audio fidelity directly into your computer, a built-in headphone amplifier for zero-latency monitoring, and simple controls for headphone volume, pattern selection, instant mute, and microphone gain located directly on the microphone.

There are no drivers to install - simply plug the Yeti into your PC or Mac, load up your favorite recording software, and record.

15 Tips for Recording Audio

While some people are recording in professional studios, even more are recording in their home office, bedroom, living room, basement or garage so we've included a few easy ways to get studio acoustic results, at home

1. **Avoid hard surfaces:** Hard surfaces can be your biggest enemy for home recording. Hardwood floors, concrete walls, tiled walls and counter tops are all reflective, bouncing sound around the room like Flubber, creating an echo that can be noticeable in your recordings.

Carpet, curtains, rounded/fabric furniture and even inside of clothes closets can help to "deaden" the room and create a more acoustically sound environment. Any hard surface is reflective; including slat blinds and windows (glass is extremely reflective). Close those curtains!

2. **Reflection shields** can be purchased or made to help minimize sounds being picked up from the back and sides of the mic (best if you are using a strictly omni-directional mic).

However, most at-home recording happens with a cardioid (forward directional) microphone. This type of mic rejects most noise from the sides and back. However, a cardioid microphone is most sensitive to sounds reflecting from behind the vocalist

3. **DIY acoustical treatment:** The best at-home treatment is to hang a thick blanket on the wall or shelf or bookcase that is directly behind the vocalist, covering as much of the surface as possible. Staying away from desks or other reflective surfaces can be beneficial.

4. **The place not to record:** in the middle of the room. This is where frequencies build up and you get "standing waves."

5. **Where to record:** Back yourself closer to the wall with the hanging blanket and further from the opposite wall. Without sticking yourself in the middle. Clothes closets can also make for great vocal booths. Clothes naturally absorb sounds; they are typically carpeted and there is less room for echo.

6. Shhhhh! Make sure that background noise is minimal or nonexistent. Turn off the AC, and the TV, and the washing machine, and the microwave, and the fan and your fish tank...well maybe just move the poor fish into another room.

Headphones and Monitors:

7. Monitors are great for playback. Headphones are great for recording and direct monitoring.

8. For at-home recording, **closed-back headphones** are best so that the sound from your headphones doesn't bleed into your microphone.

9. **Monitors** are basically fancy speakers that playback a little differently. Speakers give color to sound (deeper bass, sharper mids, etc.) and are better for the average listener of a mastered track. Monitors give you uncolored sound. This will give you an accurate sound for mixing and editing. Then use your headphones for referencing (aka listening to the audio as it will be heard in the real world).

10. **Setting up your monitors**: Set up your monitors so that you are one of the points in an equilateral triangle. Measure the distance between your monitors, example: 45 inches, then make sure that you are 45 inches from each monitor. This will be the sweet spot for listening to your monitors.

11. **Set your levels correctly**. If you are direct monitoring through headphones, be sure to set your headphone volume levels correctly before adjusting the gain. If your headphone volume is too low, you risk overcompensating with the gain and causing distortion or an overly saturated recording.

Microphone Technique

12. Popping "P" plosives: **Pop filters** help to catch these annoying bursts of air that defile your recording. But if you don't have a pop filter, angle your vocals to the side of the microphone to reduce the direct burst of air away from the most sensitive part of the mic.

13. **Distance from the mic:** The proximity effect occurs as you get closer to the microphone, the louder and more saturated your audio will become. Positioning your vocalist approximately six inches from the microphone is a good rule of thumb. Podcasters can be a little farther away, but we wouldn't recommend anything past a foot to start out. Experimenting to find your best sound is always encouraged.

14. **Pattern:** Be sure to select the appropriate pattern.

- For a single person recording, the cardioid pattern is a good default choice.
- Omnidirectional will pick up sounds from all around the room, which is great for a chorus or group, but likely not best for one person recording.

15. **Side address:** For microphones like the Yeti it is very important to remember that they are side address microphones. The microphone capsules face outward, not upward. For best results, talk into the side of the microphone as shown below.

Check out the resource section for some solid recommendations for some great inexpensive microphones that should work fine.

Audio File formats

Once you have the audio file recorded, you need to store it.

If you are going to store it on your computer, you need to choose which audio format to use.

The most popular file formats are WAV (a file format created by Microsoft) and MP3 (a compressed format used mostly on computers).

Each file format has its reported advantages, but each seems to satisfy most users' needs if the quality of the recording is sufficiently high.

You can also choose to store your audio file in a CD format, in which case you will probably play the audio from a piece of equipment other than your computer.

Today, the best choice is to store your audio in a digital format so that the quality is maintained over time, the older analog format (used on the older audio tapes) deteriorates as it is played over time.

Audio editing

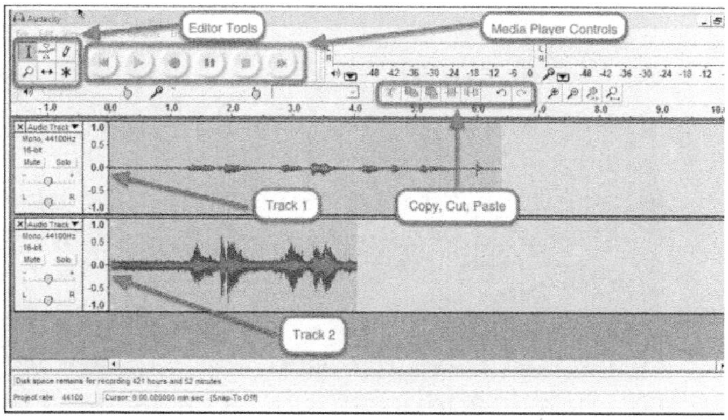

Once you have your audio recorded, you'll want to edit it for content, timing, noise removal and reduction, enhancement, and the addition of introduction, exit, and background music.

You can spend a lot of money on this software but we've found that Audacity, a free open source digital audio editor and recording computer software application, that's available for Windows and Mac, works as well as anything you can buy.

In addition to recording audio from multiple sources, Audacity can be used for post-processing of all types of audio, including podcasts by adding effects such as normalization, trimming, and fading in and out.

Audacity has also been used to record and mix entire albums and its features include the following:

- Importing and exporting of WAV and MP3 formats
- Recording and playing back sounds
- Editing via Cut, Copy and Paste (with unlimited levels of Undo)
- Multitrack mixing
- A large array of digital effects and plug-ins
- Noise removal based on sampling the noise to be removed
- Conversion of cassette tapes or records into digital tracks by automatically splitting the audio source into multiple tracks based on silences in the source material

Using Audio on PowerPoint® Slides

Although we'll explain several other audio processes for delivering information to your virtual audience, we're going to start with adding audio to PowerPoint® since it seems to be the primary vehicle that most people want to use for their material.

There are an unlimited number of advantages to adding audio to your PowerPoint® and other communication media

Since most instructors have created PowerPoints® as a backdrop for their face-to-face lectures, a logical starting point when going online is to just add audio comments to their PowerPoints® and have their audience download and view the results.

As noted, audio segments can add impact to your presentation if they are high quality segments and used properly.

First and foremost remember that any item on a presentation slide should be there for one purpose only – to increase the impact of the message being delivered.

Once you have decided that an audio segment will add to your message (and you aren't adding it just because you like it), here are some ideas to consider.

Adding sound to PowerPoint

Once you have chosen the file format, you will need to play the file during the presentation.

The most direct way is to incorporate the file into the presentation file and have the presentation software play the file either automatically or at a specified time.

To insert the file into the presentation software package, it must be compatible with the formats supported by your software.

If your file is not compatible, or you don't want to insert it into your presentation file, you will use a sound system to play your file from usually a CD or flash drive format.
This requires coordination of the presentation equipment and the sound system, but the audio equipment to play a CD or flash drive is available almost everywhere.

Either way you choose to play the audio, test the quality at a volume level that will be used for the actual presentation.

What sounds fine through headphones may sound very different when amplified to fill a large room.

If you are recording the audio segment yourself, keep these factors in mind when selecting the sound quality and consult an audio recording professional if you need further assistance.

Although you'll probably find that it's better to pre-record and edit your PowerPoint narration and then add it to the slides, there are circumstances in which it can be to your advantage to record a narration directly onto your *PowerPoint* presentation.

The **Record Narration** option lets you record your entire presentation for playback where *PowerPoint* is installed, and where speakers or a headset are available.

Tip: Save your presentation under a new name *before* beginning to experiment with the **Record**

Narration function(s). If you later decide that you *do not want* the narration after you've applied it, you'll find it easier to open a previously saved audio-free copy than to remove the audio and timing settings from the narrated version.

The process of recording a narration is simple: from the main *PowerPoint* menu, select **Slide Show | Record**

Narration

- A *Record Narration* dialog will appear.

- In the *Record Narration* dialog, you have the option to adjust input levels and recording quality or simply to click on the **OK** button and get started.

- It is usually a good idea to click the **Set Microphone Level...** button to make sure that the recording levels will give you a good quality recording.

- Upon clicking the **Set Microphone Level...** button, a *Microphone Check* dialog will appear, as illustrated below.

- Follow the instructions, reading the text into your microphone.

- As you do so, the slider (under the meter bar) will move to the right (to increase the recording level) or to the left (to decrease the recording level).

- By the time you have finished reading, the recording level should be optimally set.

- Clicking the **OK** button will subsequently close the *Microphone Check* dialog.

- When you are again looking at the *Record Narration* dialog, a click of the **OK** button, will immediately begin the recording process, so make sure that you are ready to start adding the narration to your presentation.

- When your first slide appears on the screen, the recording is underway.

- From this point, you can give the presentation as you would if you were standing before a room of people; as you make your way through the slides, the "timings" of your key-presses are linked to the audio track so that, when you've finished, your presentation is ready to be canned and delivered pretty much without you.

- If you want to pause your narration at some point during the recording, right-click on a slide to introduce the following context menu:

- Then click on the **Pause Narration** option.

- To begin recording again, right-click on the slide again, and select the **Resume Narration** option (not illustrated here).

- After you reach the last slide, you'll see a prompt asking whether you'd like to save the timings; if you're happy with the narration, click **Yes** ; otherwise, click **No** , and start the process again from the beginning.

- Once the presentation has been saved with its timings, the entire presentation will play on its own, with your narration guiding the audience and the slides advancing at the appropriate moments, in sync with your voice.

Tip: If you want to re-record a narration that has already been added to a slide, view the slide in *PowerPoint's* Normal View, and then select **Slide Show|Record Narration** from the main *PowerPoint* menu.

You will then be able to record your narration as you did before.

When you have finished re-recording the slide, advance to the next slide and then press **ESC** to stop the recording.

Software for Recording and Editing Audio

To create most your own audio files, you'll need a computer with audio editing software and a digital microphone.

Check out the recommendations in the reference section but in the meantime, if you don't already have audio recording and editing software, I recommend Audacity®.

Audacity® is free, open source, cross-platform software for recording and editing sounds

It's available for both the Windows and Macintosh platforms, allows you to record and edit, supports multiple tracks and allows you to save your work in Wav, MP3, and several other formats. http://audacity.sourceforge.net.

There's a lot of great tutorial information including tips, tricks, at http://wiki.audacityteam.org/wiki/Category:Tutorial that includes detailed information and guidance about dozens of things including:

- Audio File Formats (MP3 and others)
- Voice Effects
- Effects to improve speech recordings
- Sound Effects
- How to Apply Special Effects with Audacity
- How to Insert Background Music or Sound Effects in Audacity
- Noise removal
- How to Record and Edit a Podcast with Audacity

To get you started, here are some great royalty free background music audio tracks to embellish your production with copyright-safe music: www.jewelbeat.com/free/free-background-music.htm.

You can search thru over 4,000 free background music and sound effects simply by using keywords and there's no need to download to preview - just type in keywords, listen immediately then download what you need!

Additional delivery mediums for audio products

Because of their commercial acceptance, CDs are a great way to reproduce and offer your audio materials since just about everybody has access to a CD player.

With the drop in prices for Flash memory drives, many products are now offered in the form of a memory stick that can be played or copied onto a computer, tablet, or any device that has the USB stick input capability including many automotive players, CD players, and DVD players.

The web has created a variety of delivery techniques including sending the audio files as attachments to emails, making them playable and downloadable through websites, sharing them through cloud centers like *DropBox*™, and offering them as paid downloads through commercial venues like *iTunes*™ or from your own website as pay-per-download products.

Later on, you'll find complete details on offering your audio products as pay-per-download products from your own website. (Just like iTunes☺).

How much audio material can you get on a CD or a DVD?

First, you'll have to consider the quality of the audio recording because the higher the quality, the more space it will occupy on the CD.

As previously mentioned, the higher the sampling rate the more sound elements you'll record and will of course, correspondingly incur a larger storage space.

For recording music-quality audio the most common sampling rates are 44.1 kHz and 48 kHz and you can select them before you record.

Which Sample Rate to use

44.1 kHz (44100 Hz) is the sampling rate of audio CDs giving a 20 kHz maximum frequency. 20 kHz is the highest frequency generally audible by humans, so making 44.1 kHz the logical choice for most audio material.

48 kHz (48000 Hz) is the sample rate used for DVDs so if you are creating DVD audio discs from your Audio projects you may prefer to work with this setting.

The most common digital audio formats for recording and distribution

Most of the digital recording systems used on computers will allow you to adjust the recording frequency for best quality and then, once recorded, will allow you to save the file in a variety of formats for distribution.

WAV (.wav) - WAV is the most common format used for storing sound. Support for WAV files was built into Windows 95 making it the de facto standard for sound on PCs plus it'll play on just about every CD player on the planet.

MP3 (.mp3) - MP3 is the name of the file extension and also the name of the type of file for MPEG, audio layer 3. It compresses the audio by a factor of 12 without sacrificing sound quality and became widespread with the advent of the IPod and other MP3 players. All of the MP3 players and most of the recent CD players will play MP3.

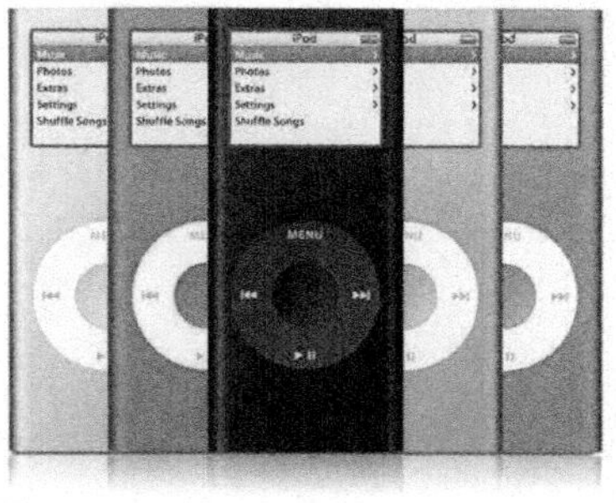

WMA - Windows Media Audio (.wma) - Short for Windows Media Audio, WMA is a Microsoft file format for encoding digital audio files similar to MP3 and will play on all Windows computer and with appropriate software, on just about any other computer, some CD players and MP3 style players but not all.

Recording and storage times

- A typical CD will hold 650 MB of data or 74 minutes of WAV file audio.
- If you store it as MP3, it will hold 888 minutes or about 15 hours.
- A blank DVD will hold 4.7 MB of data or 518 minutes (8 ½ hours) of WAV file audio.
- If you store it as MP3, it will hold 3,626 minutes or about 60 hours.
- Flash memory drives come in sizes from 500 MB to 128 Giga byte which means you can store from 60 minutes to thousands of hours of MP3.

When you store and deliver audio material from the Web, you'll want to use MP3 exclusively for its reduced storage requirements and improved download time.

And since more equipment is becoming MP3 compatible with each passing day, we strongly recommend using it as your primary, if not exclusive, recoding, storage and delivery format.

Part 4 Using Video for Training and Collaboration

What is Videoconferencing?

Here's a formal definition: *Videoconferencing is the conducting of a videoconference (also known as a video conference or video teleconference) by a set of telecommunication technologies which allow two or more locations to communicate by simultaneous two-way video and audio transmissions.*

It also includes video training and is a type of groupware.

Here's one that's less formal: *Video conferencing or video training is where one party uses a video a camera and monitor to send pictures and sound, via the internet, to another computer and monitor where the receiver is watching and listening. In some systems, the communications goes both ways.*

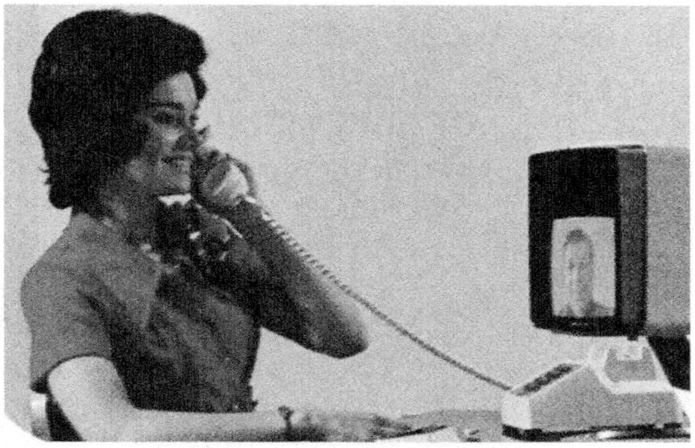

The earliest reference to this technology was the old bell Picturephone™ that millions of people saw demonstrated at Disneyland™ starting in the early 1960's.

Obviously, we've come a long way since then.

Benefits of Video Training and Conferencing

1. **Increases Productivity**

The ability to be in several places at once without leaving your office is the next frontier. The 1990s saw a tremendous increase in productivity for American companies. The goal was to do more with less. Web sites and the Internet let us quicken the speed of communication.

Most conferencing systems now have features that allow users to share and edit documents, PowerPoint, Visio, etc., in real-time.

The ability to easily share and collaborate within various formats has added another dynamic to video communications previously not possible.

Decisions are made faster, products or services are brought to market quicker, and your sales cycle can be reduced.

These benefits help increase sales, improve operational efficiency and allow businesses to stay ahead of their competitors.

2. Improves Communication & Reinforce Relationships

During a videoconference you can see the facial expressions and body language of conference participants. These are both important aspects of communication that are lost with a basic telephone call.

Videoconferencing also allows the opportunity for more of your staff to have contact with your customers. From your customer's perspective, the only person ever seen is the salesperson from your company.

3. Saves Money

An additional tangible benefit of video conferencing is businesses can start saving money by reducing travel expenses.

With a video link your employees can maintain relationships and conduct business globally without increasing your carbon footprint or spending money on airline fare, lodging and meals.

There are, however, numerous situations where you can save a lot of money on airfare and hotel costs, not to mention saving the loss in productivity from being out of the office.

In today's economy, cutting down on company travel expenses is usually a real attention grabber. In practice, the results are mixed. Individuals still like to go places and visit clients and associates in person. Nothing can replace meeting someone face to face for the first time and shaking hands.

4. Allows Multi-point Meetings Across Time Zones & International Boundaries

Videoconferencing lets you put together a meeting of various people, from different locations, for a common discussion.

Participants can be across the country or on the other side of the globe.

For example, you can have sites in London, New York and Tokyo, all with multiple personnel, participates in the same videoconference.

5. Keeps Employees Connected

Video communication helps retain valuable employees who have to relocate or need to work from home a few days.

This definitely saves on costly retraining or recruiting fees associated with hiring a new employee.

It also allows those employees who are on the road a lot to quickly keep track of new policies, developments and add insights to the day-to-day operations. Being limited by geography is a thing of the past!

5. Improve and Maintain Business Relationships

Unlike traditional conference calls, emails or texts, a live video conference allows the participants to see facial expressions and body language; meetings become a lot more personal.

Most video conferencing solutions offer recordings and storage solutions which allows businesses to save and archive important online meetings. These recorded sessions can also be used as future training tools or for compliance regulations.

6. Improve Work-Life Balance

Whether long distance or just overnight travel, excessive traveling can burn out employees and put strain on their families.

It's common to have a flight delayed or even cancelled, and traffic in most metropolitan areas is just as unpredictable. Video conferencing is more reliable and ensures your meetings take place on time.

7. Improves Effectiveness

We all know that "a picture is worth a thousand words." Videoconferencing is the perfect example of how true this statement is. A live video call is much more effective than a phone call in many situations.

For example, your engineering staff needs to show one of your vendors a problem with a purchased component. Visually seeing the part during a videoconference is far more effective and meaningful than trying to describe it over the phone.

The Benefits of Video Conferencing for Businesses

Companies of all sizes and in every industry are quickly adopting video conferencing as a way to stay connected, increase productivity and cut costs in these difficult economic times.

As prices on conferencing equipment drop and the technology becomes easier to use video conferencing is gradually becoming an efficient way for business to communicate with employees, customers and sales prospects.

Live VS On-Demand

Live Streaming vs Video On-demand (VOD)

The term "video streaming" gets thrown around a lot but it can mean two very different things.

Sometimes people are referring to Video On-Demand and other times they are referring to Live Streaming.

So what is the difference between them?

Video On-demand

- High quality (HD) video and audio
- Plays on computers and smart phones
- Plays smoothly at any Internet speed
- More economical than live streaming

Live Streaming

- No time delay
- Ability to live chat
- Ability to ask and respond to questions
- May require additional hardware and software

So which one should I use?

The diagram is designed to help you figure out which technology to use.

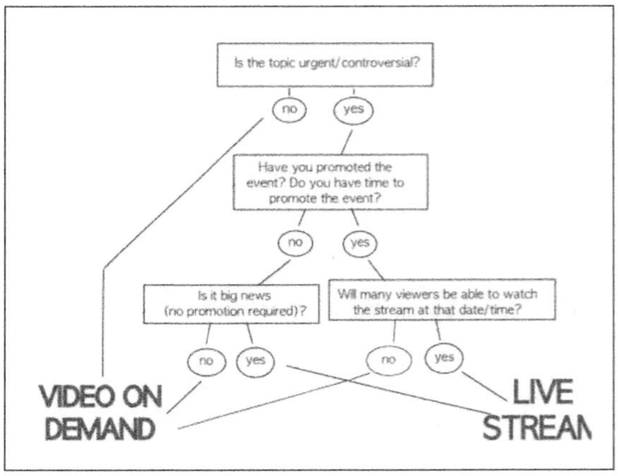

When to use Live Streaming?

If you have breaking concerns news or live events, you're probably going to want to live stream it.

If you are interested in conducting meetings and collaborating on projects, services like Ready Talk are an excellent option.

When to use Video On-Demand?

Corporate training is a good example of when video on-demand might be a better choice.

You want all of your new employees to be able to access this video, even if it's not at the same time as everyone else.

Both video on-demand and live streaming can be useful methods for your video strategy. It just depends what you are using your video for.

Preparing Quality Video Segments

The quality of the video file is starts with the resolution that you use to record the video.

Changing resolutions on video files is more difficult than changing sample rates in audio files, so you want to record at the resolution that you need for your presentation.

Most video will be recorded in a digital format and copied to your computer from the video camera or recording device and then you can use simple tools to edit your video if you need to.

Video File Formats

Digitized video can be stored in a number of different formats, the most popular being QuickTime (from Apple computer), Windows Media Video Format (WMV files from Microsoft), AVI (Audio Video Interleaved format) and MPEG (Moving Picture Experts Group compressed format).

The quality and size of the video file will be determined by the resolution you select when converting the video to a digital format.

The higher the resolution, the better the quality of the video and the larger the file size will be. You will want to select the resolution based on the ability to show a large enough picture on the screen that people can clearly see it from the back of the room.

Tips for Shooting Better Training Video with Your Camcorder

Creating good virtual training videos takes practice so we've put some tips together to get you started.

1. Go light on the zoom. Overuse of zoom is a common problem, and can result in video that looks amateurish, or even worse, can make people dizzy. If you need to zoom in and out, take it slow and steady, and don't zoom out again too soon.

2. Frame your shots well. Fill the frame with your subject, and don't be afraid to place him or her slightly off-center.

3. Include close ups especially when you're referring to something other than the person in the video.

4. Shoot from several different vantage points. Recording an entire training session from the front could be a bit dull to watch later. Footage captured from the sides as well as the front can result in a much more interesting mix of shots, and can heighten the drama of important moments.

5. Avoid backlighting. Just because we can see people's faces when they're backlit doesn't mean the camera can. Too often, backlit shots result in uncomfortably bright light haloing a dark figure with no visible features so you miss out on the great facial expressions of loved ones. You can improve a backlit situation by moving to one side or another. Some camcorders even have a feature that lets you reduce the effects of backlighting.

6. Backlighting makes a big difference in the appearance and tone of your recording

7. Unless you plan on editing, record only the important moments. If you're going to edit your footage, go ahead and record every single second of your cousin's wedding.

8. Don't overuse special effects. It's cool that you can "solarize" your footage or make it sepia-toned, or try any number of other fun recording settings. But steer clear of the fun features when you're filming important scenes. After all, if you decide later that part of the video really should have been shot in black-and-white, there's easy-to-use movie editing software that'll let you make that change.

Video lighting secrets for your studio

In most cases, the biggest difference between video perceived as being 'professional' and that perceived as being 'amateur' is the quality of lighting in the video.

A well-lit video - even one shot with a home camcorder - immediately delivers a professional image.

While a poorly lit video, no matter how expensive a camcorder was used to shoot it, says 'amateur'.

A quick an easy way to get good video in a small office is to use two banks of full spectrum, natural color fluorescent lights.

The advantages of using fluorescent lights are many and include:

- They don't product any heat, which can be a real problem when shooting in confined spaces.
- They produce natural, even, lighting that tends to surround the subject.
- They don't produce harsh shadows.
- They don't blind the talent.
- They are quite affordable, less than $40 per fixture including tubes.
- They can be left on full time.

Not just any fluorescent lights will do, though. You'll need fluorescent fixtures with electronic ballasts (this avoids flicker and hum), and you'll want to use 5000k full spectrum, natural color tubes.

Investing in a studio lighting kit can open up a range of creative possibilities and help take your video skills to the next level.

There was a time when the studio environment was largely the domain of the professional videographer.

Today, however, with a wide range of pre-packaged kits on the market, enthusiasts are finding it easier than ever before to obtain the gear and accessories necessary for creating a complete home studio.

You can pick up a basic package very cheaply that will allow you to create great looking videos or you can shell out thousands of dollars to buy a dazzlingly sophisticated and powerful set-up. There are options available for every level of experience and budget.

These types of lights do, however, have a tendency to get very warm during use (although 'cool' fluorescent varieties offer improvement in this aspect), which can make long video sessions an uncomfortable experience for your subject.

While technology varies greatly between products, you will find that many entry-level kits benefit from some of the features found in more expensive models of the same brand.

As such, it's perfectly possible to pick up a beginner-friendly kit that's still packed with features, for a modest price.

Check out the lighting kits, including the ones that include green screen components, in the resources section for recommendations.

How to use green screen to enhance your presentations

If you want to add an element of professionalism to your recorded video productions, you'll want to seriously consider using green screen technology or what is called Chroma Key effects.

Chroma key or Chroma keying, is a special effects/post-production technique for layering two images or video streams together based on color hues.

Using Chroma key, you can record the subject in front of a plain green screen and then drop in a background of your choice including stills, PowerPoint presentations and videos.

The technique has been used heavily in many fields to remove a background from the subject of a photo or video – particularly the news casting, motion picture and videogame industries.

A color range in the top layer is made transparent, revealing another image behind. The Chroma keying technique is commonly used in video production and post-production.

It's widespread usage and low cost now makes it simple to separate human subjects from the background and most commonly in TV broadcasting.

Chroma keying can be done with backgrounds of any color that are uniform and distinct, but green and blue backgrounds are more commonly used because they differ most distinctly in hue from most human skin colors consequently, no part of the subject being filmed or photographed may duplicate a color used in the background.

Commonly used for TV weather broadcasts, where the presenter appears to be standing in front of a large map, though it's actuality a large blue or green background - the weather maps are added on the parts of the image where the color is blue (or green).

The advanced state of the technology and much commercially available computer software, such as Autodesk Smoke, Final Cut Pro, Pinnacle Studio, Adobe After Effects, and now the Windows Movie Maker, makes it possible and relatively easy for the average home computer user to create videos using the "Chroma key" function with easily affordable green screen or blue screen kits.

Some Tips on Successful Chroma Keying

As mentioned earlier, Chroma Keying is a skill on its own that takes practice to do properly. When done poorly, Chroma Keying looks really low budget and obviously fake. However, there are a couple of things that even a first-time "Chroma Keyer" can do that can improve the quality of their work.

Lighting is the most important part of Chroma Keying. You can never be too careful with it. Take extra care in properly lighting your background, and use large white cards to bounce light so that you can achieve even illumination across the entire screen. Remember, different degrees of illumination can create color variations that can confuse your digital camcorder.

Also, be aware of the potential for color spill. You will see color spill as a blue reflection off of your backdrop that colors your subject.

In this case, double check the lighting on your subject (again, nice even lighting to eliminate shadows) and then have the subject move farther away from the background screen if possible.

Matte sprays can be attained from your local photography store or even art store. These are useful in reducing the reflective properties of your subject (especially if they have a bald head!) and can greatly reduce color spill or color bleeding.

Several people have figured out really cheap ways to create a green screen background but the only problem is that you need a dedicated wall to hold the green poster board panels.

Check out this video at www.youtube.com/watch?v=h8fE_1SpEWY to learn how to make one for about $10 bucks.

On the other hand, if you want a portable system or one that you can set up and take down easily, check out the resource section especially the lighting kits that include green screen components.

Sound – the most critical part of your video

The built in microphones on most camcorders are actually quite good but the farther away from the subject you place the camera (and the built-in microphone), the more ambient noise you're going to get with your recording.

Good – Max voice/least noise

For this reason, it's always best to have a camcorder with an external microphone jack so that you can use wither a wired or wireless microphone and place the microphone as close to the person speaking as possible.

This technology is changing so rapidly that it's almost impossible to recommend anything except to take a trip to BestBuy™ and see what's currently available in the way of cameras with external audio jacks.

Check out the article at http://www.camcorderwithmicrophoneinput.com for more details including video links.

As far as microphones are concerned, you can find full range of good quality microphones at reasonable prices at places like the Guitar Center (www.guitarcenter.com)

Tips for creating good video podcasts:

- Do not use the wide-screen mode but use the standard mode instead.
- Optimize bandwidth and battery power by shooting several short segments instead of one full-length film.
- Small screen viewing will not show text clearly unless it's large enough.
- Fast motion and high contrast might not show very clearly on the small screen.

Video Editing Software

After you've shot your video, you'll want to edit it and add in special effects and other goodies to enhance the final version.

The fundamentals of consumer video editing software haven't changed much over the past few years:

You want a program that lets you import video, stills, and sound to your PC from whatever camera or source you have, easily join and trim what you shot, and maybe add some transitions and cool effects.

If that's all you need to do, you can rely on the simple video editing software that comes with both Windows—the much-improved Windows Movie Maker, and with Apple Macintosh computers' iMovie app.

You probably have the Microsoft or Apple software installed on your computer but if you don't, just visit the Web sites listed below to download them for free.

Microsoft Movie Maker for Windows users and Apple iMovie for Mac users are probably the two easiest to use free video editing software programs available.

Wax and Lightworks are high performance, free video editing software programs that're good for both home users and professionals.

They can be used as a stand-alone application, or as a plug-in to other video editors. The software also features unlimited video and audio tracks.

- http://windows.microsoft.com/en-US/Windows-Live/movie-maker#t1=overview
- www.apple.com/ilife/imovie
- www.debugmode.com/wax
- http://www.techradar.com/downloads/lightworks

If you're willing to spend $100 bucks or less, for some additional features, then I'm going to recommend the following packages.

Adobe Premiere Elements 11

$99.99 list - $59.99 at Amazon
The latest version of Adobe's consumer video editing software gets a welcome simplifying makeover. Quick and Expert modes let you choose the level of interface right for the task, and a new project assets panel helps you get organized. Filmlooks effects can give your movie a Hollywood feel, but the program still lacks support for 3D and 4K video content, and slows down noticeably when you apply some advanced effects. Premiere Elements is the only non-Apple product here that also runs on Mac OS X

Corel VideoStudio Pro X6
$79.99 - $59.99 at Amazon Corel VideoStudio Pro X6 is a surprisingly power-packed yet easy-to-use video editor. It leads the pack in some things like stop-motion capturing and motion tracking effects. The interface is clear and easy to use, and the program performs quickly and stably. Included is a screen-cam utility, support for the new 4K Ultra HD video format, and the ability to create HTML5 web video code.

CyberLink PowerDirector 11
$99.99 – List - $74.99 at CyberLink
CyberLink continues to impress with its high-power enthusiast video editing software, with cutting edge capabilities like support for Ultra High Definition (4K) and 3D video formats. Despite its wealth of powerful effects and tools, the program interface is very clear and easy to use. It's one of the only products that doesn't slow down on a midrange tower PC when performing demanding multi-track overlays. Direct sharing to the popular video websites and strong audio and disc authoring capabilities round out the package.

Magix Movie Edit Pro 2013 Plus
$99.99 - $72.95 at Amazon Magix has been in the video editing software game for more than 20 years, and the experience shows. The attractive, easy-to-figure-out interface belies a ton of powerful, near-pro-level tools. You also get screen-cam capability and HTML5 video code output, and strong music and audio editing tools. Magix only slightly trails other offerings such as PowerDirector in helpful tools, speed, and support for new formats like 4K Ultra HD video.

Pinnacle Studio 16 Ultimate
$129.99 - $65.88 at Amazon Corel's purchase of Pinnacle from Avid resulted in this full-featured, fast-performing nonlinear video editing program, actually derived from the discontinued Avid Studio. Pinnacle Studio has an attractive interface design, tons of video effects including a nice stop-motion tool, and 3D capability. It lets you directly upload to Facebook, Box, YouTube, and Vimeo. Its disc authoring tools offer lots of

menu templates, but it doesn't yet offer 4K support, and some procedures are less intuitive than in other programs.

Sony Movie Studio Platinum 12
$99.95 - $34.99 at Amazon Sony Movie Studio HD Platinum 11 offers 3D editing, accelerated performance, and near-professional-level control, but at the cost of near-pro-level degree of difficulty. The software doesn't yet support 4K Ultra HD video content, suffered occasional crashed during testing, and doesn't offer picture-in-picture preset like those found in competitors. It also limits you to 20 simultaneous video tracks, compared with the 99- or 100-track limit in most competitive packages.

Video formats for downloading

Note: Flash doesn't work on tablets or smart phones

Although Flash® videos are all over the place and will play just fine on your Windows™ or Macintosh® computer; Adobe was very public about dropping mobile Flash and has already said that HTML5 was the way forward on phones and tablets.

The bottom line here is that not all videos can play on all types of devices.

That being said, MP4 is currently the standard video format for smart phones, and tablets plus they'll play on your Windows or Macintosh computer.

That's because MP4 is the jack-of-all-trades among video formats. It's widely used for computers and mobile devices, and it's appropriate for both downloading and streaming video.

Since it's the default choice for smartphones and tablets, we'll recommend using MP4.

AVI Format

While MP4 is the recommended standard, an old standby format is still widely used.

Just about every media player supports AVI movies, and you'll often find AVI videos on sites that offer free movie downloads.

Developed in the early 1990s, AVI has become somewhat antiquated since there have been several technologies introduced that offer better compression schemes for optimizing space requirements.

That's why AVI files are typically larger than MP4 files, making them <u>less suitable for use on mobile devices</u> or for streaming movies on a computer.

If you have a library of AVI files on your computer that you'd like to offer as digital products, there are plenty of video format converters available that will do the trick.

Part 5 - Video Products and Podcasting

When the first talking movie was released in 1927, the world of audio-visual content changed forever.

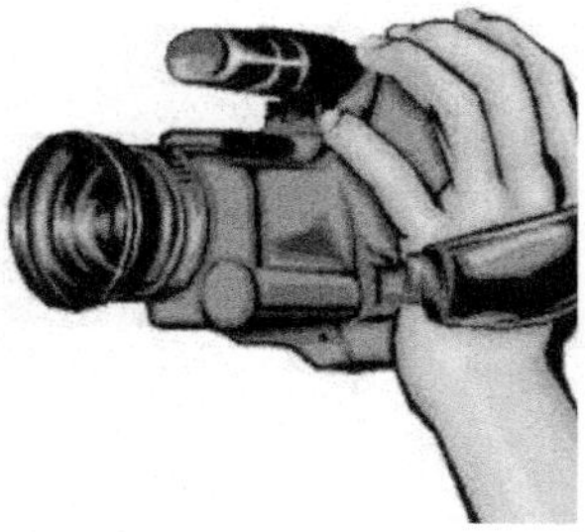

In the 1960s, reel-to reel video tape made its debut and in the 1970s, video cassettes began to dominate the scene.

In the 1990s, DVDs appeared and in the new millennium, the ability to generate video from your Smart Phone and a pen in your pocket coupled with the ability to distribute the video through places like YouTube™ and Netflix® have made video as commonplace as the daily newspaper once was.

Why use video?

In the section on audio, I got on my soapbox and extolled the virtues of audio products over video products and, although my logic was sound, there are still a lot of places where audio MUST be replaced with video especially, if the subject matter mandates that there be a visual component to guarantee that the subject matter will be understood properly.

The advantages of visual communication are:

- Facilitating quick understanding of complex communication by using pictures and symbols
- The receiver gets the idea fast. A picture is still worth 1,000 words and visual communication does not require any further explanation for the receiver or by the sender.
- Also it enables viewers to understand things at a glance: When the logos or trademark are posted, it will be easily understood to the people.
- The receiver gets the message more clearly with than any the other media of communication.
- The visual component makes it easy for the recipient to make a comparison since a consumer's buying decision is usually accompanied by a visualization of the product compared to the competitor's product. Visual media can be analyzed again and again and facilitates the comparison.

- This facilitates to memorize things. Visual communication is a good tool to memorize things. It is a good way to teach kids with the help of visual pictures.

- The same can be done in marketing activities. Visual communication becomes a bird's eye. That is a bird watches the world on a tree can see every angles and it can wrist its neck in different way. The same application is possible for a consumer to see things in different way.

However, no approach is perfect and a prime disadvantage of visual communication is that the cost of going visual communication is very high.

- There should be a high consumption of time and needs experts to be appointed. Preparation of visual communication objects takes lot of hard work and dedication.
- Visual media alone is inefficient. It is done with the help of other media such as audio, computer based media etc.
- The visual media lacks the content. It can give the idea. But cannot tell the whole story.
- If the sender wants something to add other media support is necessary. Designing a good visual communication is very difficult.
- Sometimes it may lead to failure.

So most successful organizations opt for a combination audio and visual based communication.

Let's start with the benefits to the world of corporate training.

Corporate training programs often utilize various mediums to increase the effectiveness of training, and video training is one of them.

Video training can be used as a stand-alone tool with no trainer present, or can be a supplemental aid with a live trainer to enhance interactive training sessions.

Portability

Video is a portable medium, requiring only a playback device and a viewing screen.

Most trainers utilize a portable projector and laptop for larger groups, but a laptop alone will suffice for individuals and smaller audiences.

This allows the training facilitator to travel to different locations and quickly set up and break down equipment, or convert an office or other room into an instant training facility.

Consistency

Consistency of content is another major benefit of video training. Especially when no trainer is present, all viewers receive the same content.

When used in facilitated interactive sessions, the trainer is able to comment and clarify points within the video presentation and allow attendees to ask pertinent questions.

In large organizations where different trainers facilitate the same course, video ensures that the same content is presented despite differing personal training styles.

Cost Effective

Using video enables you to better plan and predict training session times, saving you and your employees valuable work time.

Costs to produce your own videos will vary, but you may purchase, rent or lease video trainings if production costs are not in your budget.

Video can cut training times considerably when used properly, particularly if the training material doesn't require a live facilitator.

Detailed Modular Information

Video is able to present large amounts of detailed information in a finite, condensed form.

If produced properly, video can eliminate or reduce confusion and questions that sometimes arise when written or spoken information is used during presentations.

Video can also be modular in content, allowing further control over the content and timing of training sessions.

A well-planned training program using video will employ scheduled employee sessions, presenting them with information pertaining to their specific job on an as-needed basis or as part of a comprehensive program.

Fundamental Uses for Video in Business

1. **The Customer**

Nothing is more compelling than seeing and hearing real customers sing the praises of your products and services.

We do that with video customer testimonials, success stories, documentary case studies, and man-in-the-street Interviews.

Use your creativity and experience to help you do this well.

Then there is market research, focus groups and polling — think of what you could learn.

2. **The Product or Service**

- Product demos show how your product works and highlight the features that differentiate it from that of your competitors.
- Product presentations are when you let your customers tell the visual story of your product or service.
- Third party product reviews are trusted by your customers. Could you be using all three?
- In Store Video and company lobby or waiting room video?
- Corporate Communications Videos
- Corporate overviews, executive presentations & staff presentations — video is a great way to get these messages out there. Recruitment Videos

3. **Training and Support Videos**

Training — the no brainer of business video! If your team is all over the country, state, or world — expert videos for sales support can fire them up at once.

Or save thousands of dollars with post sale support and maintenance videos.

4. **Internal Communications Videos**

Video is effective for internal communications, change management, employee orientation, compliance, health, legal & safety.

Highlight business plans, new business activities, and direction

5. **Advertising , Marketing and Promotion**

Even if you don't use broadcast commercials, there are ways to advertise using online video pre-roll, online sponsorships, in-game advertising, event sponsorships and in-theatre advertising.

You can use viral video, or video in landing pages and micro sites.

6. **PR Support and Community Relations**

Beyond the video press release, there are news packages & other PR support materials.

If your company is helping the environment or contributing to important causes you should be capturing those efforts on video for your community relations reel and corporate responsibility video

7. **Event Video**

Why limit the event to the size of the venue?

These days, event presentation video get thousands of extra views, round table sessions go viral, live internet broadcasting open up your expert Q&A sessions to the world.

8. **What about traditional education?**

Everywhere you look on the Internet these days, you can find video. And lots of it!

In October 2009, Google reported that YouTube alone was delivering one billion views per day.

Video is a medium that engages viewers from multiple senses – sight and sound – and can generate excitement about a subject or concept.

Students will enjoy the experience and retain more information from the class.

So, it's no wonder that more and more faculty are looking into using video in their instruction. Here are just a few ideas of what you can do with video:

- Course welcome
- Icebreakers
- Launching point for group discussion
- Mini-lectures
- Simulations/demonstrations
- Student projects/student-created media
- Feedback for students
- How-to videos (great for teaching software or a procedure)

As with any instructional technology, you want to use video to enhance teaching and learning.

To effectively integrate video into your course, you must first determine a specific learning objective and create an activity that uses the video in support of that objective.

Begin by asking yourself these questions:

- What is it that you want your students to learn?
- Is the video to introduce new concepts, review old ones, or extend something that happened in class?
- Is it to provoke thoughts and promote critical thinking, or is it to provide simple, factual information?

The overall benefits of offering video products

Now that we've addressed the world of business and education, let's examine the world of entertainment.

You're probably not going to be competing with the major networks for market share, but that doesn't mean that you can't or won't be offering a lot of material that's both interesting and useful.

Social media is becoming a standard channel for instant communication, online marketing and sharing of knowledge.

Whether you're a fan or not, you really can't ignore it. We are all becoming a part of it.

One increasingly popular part of social media is online video. (can you say YouTube™ boys and girls?)

Not only is video being used for marketing, but it's also becoming a common method of communicating and sharing.

If you're on the fence about using video in your business or on your blog, here is a compilation of some of the benefits that may convince you to get in front of the camera and give it a try:

- You can build a rapport with your audience.
- You will instantly become more personable.
- It can be a powerful way to promote your brand.
- You can motivate potential clients to act quicker.
- If your competitors do it and you don't, you could be losing clients.
- You don't have to type anything for a change.
- You can demonstrate your expertise.
- You can show your creativity.
- You can have fun.
- You can show your human side by posting a not-quite-perfect video.
- It may be easier for you to retain the attention of your target audience.
- You can develop a level of trust beyond what's possible in writing alone.
- It's an inexpensive marketing approach.
- You'll show that you're keeping up with technology.
- You can experience increased visibility in search engines.
- You can give your audience a way to benefit from your knowledge passively (it's a lot easier to watch a video than read a screen of text!).
- You can provide targeted training.
- You'll become more memorable.
- You can "show" instead of "tell."
- You can reach more of your target audience.
- It can help you establish yourself as an expert.
- You can get people to come back to your site more frequently.
- You can tap into the emotions of your audience.
- You'll make it easy for your viewers to pass on your message.

- You can use your videos to make announcements, share news and offer special deals.
- It's entertaining for your viewers.

Back to PowerPoint®

Since PowerPoint now allows for the embedding and linking of video clips, should you be using them?

The answer is that different scenarios will offer different results when choosing to link to or embed a video into a PowerPoint presentation.

First, you need to consider that PowerPoint has come a long way with regard to adding video into a presentation.

Now you can embed a video file that you have saved on your computer, or you can link to a video on an internet site (such as YouTube) by embedding the HTML code onto the slide, rather than the video file.

Embedding and Linking Video in PowerPoint®

There's more than one way to add a clip to a slide so let's look at both embedding and hyper linking your videos.

Embedding a video clip

Video can be embedded into a PowerPoint presentation.

This means that it is actually a part of presentation and is built into the slide you embed it into.

This is helpful since it means you will not forget to copy your video from your CD or log on to the Internet and realize your video is not playing until you are in the middle of your presentation.

Remember, your video must be .MPEG or .AVI to embed. To embed your video, follow the instructions below:

Open up your PowerPoint story in slide view, display the slide you want to add sound or movie to.

- Go to the Insert menu click **Insert.**
- Choose **Movies and Sounds** from the cascading menu.
- You can either choose the Movie From Gallery option or the Movie From File option choose **Movie from File**.
- It will then ask you where your movie is.
- Navigate through your files until you find the folder that contains the sound or the movie, and then double-click the movie you want.
- An icon will appear on the slide.
- PowerPoint will ask you if you want the movie to play automatically.
- If you want the movie or video clip to start playing when the slide comes up you will need to click Yes or No.
- If you select Yes, it will play once you click to the slide in your presentation.
- If you select No then you can choose when you would like the video to play or you can manually click on the image for the video when you want it to start playing.
- If you insert more than one video clip on a slide and you said yes to play automatically the video clips will play in the order that you inserted them on the slide.
- The first frame of your movie will appear on your slide.
- The video clip icon, a blacked-out video screen, will appear on your slide.

You will know that it is a movie because a movie strip symbol appears in the frame. There will also be a movie toolbar that you can use to look at your movie and make it loop.

Remember that looking at your movie in the presentation format is different—and is what people ultimately will see.

Hyperlinking to a video clip

Create a hyperlink to a sound file you want to play.

1. Select the object you're linking to, and then choose **Insert | Hyperlink**.

2. Click the **Existing File** or **Web Page** option at the left of the

dialog box, and then click the **Browse for File** button at the right.

3. Find the file you want to link, select it, and click **OK**.

Note: Linked video clips are not part of the slide show, so you have to make sure the file is available (i.e., on your local hard drive) when you play the slide show.

Playing the video in your presentation

By default, the video runs when you click on it during a slide show. You may not want this; therefore, after you've inserted the video on the desired slide, while still in slide view, you need to set the options for the video to play during the slide show. To do so, follow these steps:

1. Select the video icon and choose **Slide Show/Custom Animation** to open the Custom Animation dialog box (or right-click to access the shortcut menu).
2. Click on the **Play Settings** tab and select the options you want.
3. For additional choices, click the **More Options** button. For example, you might choose **the Play Using Animation Order** and **Pause Slide Show** (while playing) options. You can also click the **More Options** button and select **Rewind Movie**
4. **When Done Playing** from the **Movie And Sound Options** panel in the Play Options dialog box.
5. Click **OK**.

How to have video play on a mouse click

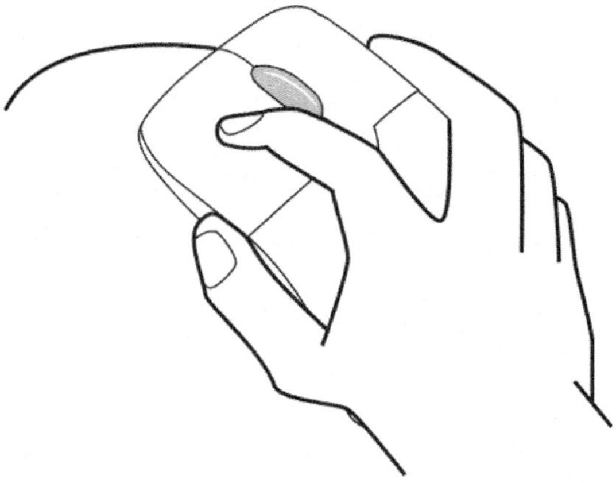

You can set your movie to play on a mouse click or automatically.

To set this up, go back to the Slide Show menu.

1. Choose Custom Animation.
2. Choose between on Mouse Click or Automatically in the Timing Menu.

How to have video disappear after playing

You may want your movie to disappear after it has finished playing.

1. Go to the **Slide Show menu**
2. Choose **Custom Animation**.
3. Choose the **Effects** menu.
4. Choose **Hide After Animation**.

How to adjust the size of your video

If you want to adjust the size of the video, you can resize it as you would any other PowerPoint object.

1. While in Slide view, click the video to select it.
2. Choose **Format**.
3. Choose **Picture**.
4. Next, click on the **Size** tab.
5. Select the **Best Scale For Slide Show** option, and then click OK.
6. When you select this check box, it prevents the video from skipping while playing.

You can also adjust the size of the video manually.

Simply select it and drag one of the sizing handles (your insertion point changes to a double arrow) and pull it in the desired direction.

If you want the video to take up the entire slide, drag the sizing handles accordingly.

Playing Video in the Presentation

- Make sure that the size of the video is large enough so people at the back of the room can clearly see the video – test this so you know for sure.

- You will need to send the audio part of the video segment to the sound system so that people can hear the audio part as well as see the video part.

- As with an audio segment, test the volume of the audio part of the video segment to make sure everyone will be able to hear it clearly.

Advantages and disadvantages of visual communication

Our modern world becomes more visualized in every aspect due to the high influence of media.

A consumer who thinks about buying a product will definitely visualize it in the mind for a while, so the use of visual communication is crucial to business success.

Visual communication is the communication in which pictures, colors, graphics are used. A skull and two cross bones shows the meaning of danger.

The advantages of visual communication are:

- Facilitating quick understanding of complex communication by using pictures and symbols
- The receiver gets the idea fast. A picture is still worth 1,000 words and visual communication does not require any further explanation for the receiver or by the sender.
- Also it enables viewers to understand things at a glance: When the logos or trademark are posted, it will be easily understood to the people.
- The receiver gets the message more clearly with than any the other media of communication.
- The visual component makes it easy for the recipient to make a comparison since a consumer's buying decision is usually accompanied by a visualization of the product compared to the competitor's product. Visual media can be analyzed again and again and facilitates the comparison.

- This facilitates to memorize things. Visual communication is a good tool to memorize things. It is a good way to teach kids with the help of visual pictures.
- The same can be done in marketing activities. Visual communication becomes a bird's eye. That is a bird watches the world on a tree can see every angles and it can wrist its neck in different way. The same application is possible for a consumer to see things in different way.

However, no approach is perfect and a prime disadvantage of visual communication is that the cost of going visual communication is very high.

- There should be a high consumption of time and needs experts to be appointed. Preparation of visual communication objects takes lot of hard work and dedication.
- Visual media alone is inefficient. It is done with the help of other media such as audio, computer based media etc.
- The visual media lacks the content. It can give the idea. But cannot tell the whole story.
- If the sender wants something to add other media support is necessary. Designing a good visual communication is very difficult.
- Sometimes it may lead to failure.

So most successful organizations opt for a combination audio and visual based communication.

Using Online Video to Drive Traffic

In the late 1990s and early 2000s when the Internet started heating up, a lot of experts got on their soapboxes and predicted that television would be slain by this new monster called the World Wide Web.

Even today, these so called experts see the Internet as becoming the predominant form of "information receiving" and that TV will soon be a thing of the past. For this reason, video will continue to be the #1 attention grabber for years and years to come.

So the question becomes: Is the Web a good place to advertise and drive hot traffic to my website?

Yes! Yes! And yes! As for online video, there are some jaw-dropping statistics, including the fact that YouTube reported 4 billion DAILY global online video views in 2011. Therefore, it's important that you ride the wave online and start producing videos for your business.

Here are the top 4 most effective uses of online video to drive traffic:

1. Testimonials: Testimonials are the absolutely crucial if you're interested in transforming your business and dramatically increasing sales conversions. A great way to use video is to record testimonials of your clients saying warm and fuzzy statements about your business, your products and your services.

   Preferably, the video would be face-to-face with customer, but you can also do it virtually using Skype and a nifty little add on called Pamela for Skype, which lets you record the audio and video of any Skype call. Once you have the videos, plaster them prominently on your website, send out email blasts to your email distribution list, and post it all over your social networks. People will instinctively trust your company more when they see and hear how other customers found your product/service useful.

2. Interviews: This is a GOLD MINE! I'm really surprised that more businesses don't do it. No matter what industry you're in, there are literally hundreds of executives, experts and household names that have a wealth of knowledge on your subject matter.

Why not conduct and record simple interviews with these folks to drive traffic to your website and social networks? You'll probably have to send out quite a few requests for every interview landed, but if you approach it correctly, you may find yourself on a Skype call with some industry heavy-hitters. When people see that you've interviewed these big name individuals, they immediately become intrigued and associate a higher degree of credibility with your company.

3. How-To Videos: This should be pretty self-explanatory to most folks. The idea is to record videos that help your audience (customers, potential customers and referrers) solve common problems and overcome obstacles that relate to your industry.

For instance, an accounting firm may do a tutorial on using Quickbooks Online. Providing these valuable tips solidifies you as an expert in your industry and gives you a ton of exposure on networks like YouTube.

4. Events: These opportunities don't come around quite as much as the others, but when they do, it provides a very good source of content for your audience to consume. Consider recording pivotal moments at an industry tradeshow or even getting video at an awards ceremony in which someone in your organization is recognized.

Remember, this is just other way to keep your audience engaged and your brand at the top of their minds.

Each of these are great opportunities to keep all eyes on your brand. Don't worry about creating a Hollywood blockbuster, just grab a decent camera and start recording. Once done, share your videos with the world by posting to YouTube.

If you use this extremely powerful, low cost means of marketing, you're sure to reap huge dividends!

Part 6 - The Current Technologies

When you refine all of the technology available for virtual communication and training, you're left with three distinct processes and a whole bunch of different virtual information technology packages for delivering them.

In this section we're going to examine the multi-media parts, audio and video, and leave the support materials for the next section because they have a lot of options and process all by themselves.

We're going to make some solid recommendations for hardware, software, and delivery systems based on three factors:

- Ease of learning and usage
- Cost
- Performance

Wherever possible, these will be based on tests we have conducted or equipment and software that we actually use for our own presentations.

Here are the three areas with technologies that we'll cover:

1. **Audio**
   - Live feed audio
   - Audio podcasts
   - Teleseminars
   - Teleseminars with interaction
2. **Video (with audio included)**
   - Live video feeds
   - Video e-mail
   - Video podcasts
   - Webinars
   - Webinars with interaction
3. **Support materials (Next section)**
   - Books and manuals
   - PowerPoint Slide shows
   - Software
   - Databases

Preparing for a Virtual Presentation

Before you decide to use one of the technology tools that are available to you, you need to remember that the key to a successful virtual presentation is the same as the key to a successful face-to-face presentation: PREPARATION!.

Consider the following guidelines to ensure that you're ready for your virtual presentation.

Knowing which non-virtual rules apply

Always remember that a virtual presentation is still a presentation. The following basic rules for making an effective presentation apply:

Identify the purpose for the presentation.

Make sure you know why you're presenting. No one wants to hear a presentation, virtual or otherwise, that has no real purpose. Identifying the purpose of your presentation is critical for knowing how to construct it and deliver it.

Know your audience. Find out as much as possible about your audience.

1. Organize your presentation. Choose a pattern that suits your topic (for instance problem and solution; past, present, and future; theory and practice; or whatever works). Make sure that your audience can easily follow your message.

2. Write an introduction and conclusion. Just because a presentation is virtual doesn't mean it doesn't need an opening and a close. The introduction has to preview and lead into your talk. The conclusion has to wrap it up.

3. Provide materials in advance. This allows audience members to become familiar with the proceedings beforehand and get more out of them. You can easily distribute materials via e-mail. Consider giving out the following:

   - Agenda: The audience wants to know what will happen and when. That's the purpose of an agenda. So provide one. An agenda makes it easier for audience members to keep track of the presentations.

   - Support material: Providing support materials such as presentation notes and outlines in advance also helps attendees get more out of your presentation. These can be sent easily and cheaply through e-mail.

  - PowerPoint slides: For presentations given on the Web, it may be smarter to distribute PowerPoint slides beforehand rather than using PowerPoint during your presentation.

  - Contact information: A list of all audience members' names, phone numbers, and e-mail addresses should be distributed so that they can reach one another in the event of a technical problem.

- Check the room(s). It's standard advice to check the room in advance of a presentation. You want to make sure that it meets your needs: Check to be sure that it has enough seats for the people participating in the meeting, that the room layout is conducive to discussion, and that the equipment you plan to use is in working order. In the case of a virtual presentation, you won't be able to visually inspect all the meeting rooms, but do check the one that you'll be presenting from.

Considering unique factors

Because of the differences between virtual presentations and face-to-face presentations, you need to consider the following special factors:

1. Timing: Unlike an in-person presentation, a virtual presentation can span several time zones. You need to take this into consideration when scheduling the presentation. If possible, give the presentation during normal business hours for all audience members. Be aware of when audience members at various locations usually take breaks or eat meals. Try to plan a presentation time that is convenient for all of them.

If you have a regularly scheduled virtual meeting (weekly, monthly, etc.) that includes people from several time zones, rotate the starting time. That way, no one has to get up early or stay up late every time.

2. Technology: Make sure that all audience members have the equipment necessary to receive your presentation and that they know how to use it.

  - Test the equipment beforehand. Then test it again. If the equipment doesn't work, your presentation doesn't work. Have a backup plan ready in case the equipment fails during your presentation. (Will you call everyone? E-mail everyone? Shout out the window?) If audience members are using a phone to connect with others, they need to make sure the line doesn't have the call-waiting feature or that call waiting is temporarily disengaged.

3. Appearance: If you appear on screen via a Web or videoconference, then you should consider how your appearance will affect your message. Here are a few ways to improve your electronic appearance:

- Wear a conservative outfit that doesn't draw attention away from your face.
- Wear pastel colors rather than bright colors, because pastels broadcast better.
- Avoid striped, checked, or patterned clothing.
- Make sure your clothes contrast with your background colors.
- Use a solid-color background.
- Remove objects from the background that may distract your audience, such as pictures, posters, and so on.
- If you're doing a green screen video, make sure:
    - That everybody's clothing and accessories are not going to be a problem with the green screen
    - That the lighting and positioning of the set, green screen background, and people in the shot are all positioned so that the green screen effect works properly.

Live Audio

Live audio is just what the name implies, you and your guests will be performing live for a virtual audience and you'll be the host of you own radio show albeit web rather than airwave based.

If you really get into this, you can carry a small digital audio recorder around with you to interview people, do on-site narration of events, and edit and combine it all together to create your own radio show.

Although you can get a whole bunch of free stuff to play with, remember that you're going to be learning software processes that you'll have to unlearn and abandon if you're serious about this method of delivery.

There are other problems with some of the free and low cost (cheap) systems like limited listeners (6, 12, or 20). If the number of listeners exceeds the pre-set numbers you'll have to upgrade to their full package anyhow so what the heck☺)

Rather than have you waste your time and money, if you're serious about this method of delivery I'm going to recommend that you invest $129 bucks for BroadWave because it offers two different options for audio streaming

1. You can stream live audio instantly over the Internet
2. You can Stream pre-recorded audio from saved files

They do have a test package to try their audio streaming server system. Seeing and using the software first hand can answer most questions

It has a whole bunch of great features that make it easy for you to use:

1. It runs on any PC as a server to stream live audio over the web.
2. It can also stream pre-recorded audio.
3. Its content does not need any special software to listen to a webcast.
4. It streams and plays directly from the default web browser.
5. The professional version does NOT limit the number of listeners

Here's a bonus: You can use it for FREE if you add a link.

- You can use the full version of BroadWave free if you link to them on your site.
- If you cannot link to us you will need to purchase after 14 days.

Here are some typical streaming audio applications for you to consider

- Stream live or pre-recorded company announcements
- Broadcast your own online radio program
- Air live music concerts or events to a worldwide audience
- Create and share your own Internet podcasts
- Reach a wider audience by streaming religious services
- Internet radio broadcasting and podcasting

For a lot of 'old timers', this is audio nirvana because it allows you to have your own radio show with or without, pre-recorded segments.

Move over NPR Radio!

I've used several of the NCH software packages and love them. They're comprehensive, easy to learn and use and they provide constant improvements and upgrades at no charge.

They also have related software packages that integrate perfectly with each other for recording, editing, and other special needs.

This is NOT a regular Podcast system where you can sell subscriptions and monitor listeners, and although you can set up your web site to sell subscriptions, this is essentially a FREE (for the listener) system where they can go to your web site and listen to your show whenever they want to.

Since you can save the shows, you can have an entire series stored and available on your site to support your listener's needs.

Podcast your material

Have you ever wanted to be the star of your own radio show?

Well now you can be using a medium known as podcasting.

A podcast is a radio show that is made available to listeners via the Web.

Just like a radio show, a podcast consists of a series of individual episodes that can be listened to on your PC or MP3 player.

Episodes can be listened to one at a time, or visitors can subscribe to the entire series using software such as iTunes.

Podcast Definition Summary

Podcasts are:

- A new type of media that puts you in control of what you listen to, when you listen to it and how you listen to it (usually on a computer or portable audio player)
- An easy way to share your own audio content on the internet with a focused audience
- A media that can be created with equipment you probably already own, for little cost and in as little as a day
- A way to become a recognized expert or celebrity in your niche
- A powerful way to connect with your customers and promote your business to them on regular basis
- Podcasts can be a great way to market your business.
- Podcasts are short audio files that can be used to easily share information with others through your business website or via distribution tools like iTunes.

- Podcasts are generally available to users at no cost and represent a way for businesses to share expertise that can serve to market their products and services.

No Professional Equipment Required

One of the big advantages of podcasting is that it does not require any special equipment.

You can easily create your own podcasts by using the audio functions found on most computers.

This advantage can also create a disadvantage, however, in that sometimes the sound quality of the recordings may not be exceptional; depending on your intent with these podcasts, that may reflect negatively on your message.

You can improve the sound quality of your recordings by investing in a dedicated microphone for your computer or a good headset so review the information above and the recommendations in the resources section.

One of its chief advantages is that the podcasts are stored so the listeners can download and listen to them when it's convenient for them.

Imagine getting new "radio"-style talk and music shows to listen to on your iPod or other MP3 player every day. You wake up and automatically have new shows ready to listen to while you exercise or commute to work. This is the podcast listening experience.

Not only that, but anyone can create his or her own "radio"-style show and broadcast it to the world in very little time and at very little cost. All you need is a computer, a microphone and the internet.

Why a Podcast Makes Sense

Let's face it. Broadcast radio is getting old. The same songs get played over and over until you're sick of them. A few large companies have control of the formats and what gets played. There is no variety.

If I listen to the radio while driving, one hand is on the "scan" button and the other is on the wheel. As soon as something loses my interest, I tap the button—and I tap it a lot.

With podcasts you don't have to wait to hear something that interests you. You choose what you want to listen to. There are now thousands of podcasts and the number is growing rapidly.

Podcasting has leveled the playing field. Anyone with something to say and a desire to say it can create a podcast and publish it to the world.

Podcasters (as they are called) are not restricted by the FCC or traditional broadcast formats and regulations. This allows for numerous subjects and formats.

Podcasting allows you as the listener, and you, as the broadcaster, to decide what is worth listening too and make it available.

More About What a Podcast Really Is

Podcasting stands for Portable on Demand Broadcasting.

A podcast is like a radio show. However, instead of being broadcast live, a podcast is

A recorded multimedia file that is distributed by subscription (paid or unpaid) over the Internet, so that you can listen to it whenever you please.

It can also be

- Played back on mobile devices (iPod, MP3 players, iPhone) and personal computers
- Podcasters' Web sites may offer direct download or streaming of their content
- May be downloaded automatically through iTunes store
- The word podcast is a play on the word broadcast combined with the word iPod (one of the most popular portable MP3 players).

Apple did not invent the word podcast, although I'm sure they are enjoying the free marketing they are getting from the popularity of podcasting.

You do not have to have an iPod to listen to a podcast. In fact, there are many other MP3 players that are better for listening to podcasts.

You can listen to podcasts on anything that plays MP3s. Once the podcast is downloaded, it's just an MP3.

The Podcast Revolution: You Can Create or Listen to Whatever Kind of Audio Content You Can Imagine.

Anyone can create a podcast. All over the world, people are creating podcasts on subjects ranging from movies, to technology, to music, to politics and whatever else you can think of.

This is new original content made by passionate people who want to share their creativity with the world.

The cost to start podcasting is so low that anyone can do it. Most podcasters are everyday people like you and me.

They could be talking to you driving in their car, sitting in their living room or speaking at a conference. You get to glimpse into their life and into their interests.

Podcasters are creating very raw and real content and listeners are responding.

Free from corporate radio and broadcast regulations, you can create whatever kind of show you can imagine.

Some podcasts are "talk show" style. Others introduce you to the latest bands and music.

With podcasts you can stay current on the news, get a glimpse into someone's life, listen to move reviews and the list goes on.

Most podcasts are made by people who are very passionate about their subject. Passion is infectious and interesting to listen to.

Since the birth of podcasting, a huge variety of shows have shown up on the internet.

In addition, many major media outlets are also now offering podcasts.

You can listen to NPR, Rush Limbaugh and many major news outlets by podcast.

What Makes Podcasts Different and So Popular to Listen to?

Maybe you're still wondering what all the hype is about.

MP3s have been on the internet for quite some time. Internet radio is nothing new. Audio blogs have been around for a while, too.

What sets podcasts apart is that they can be automatically downloaded to your computer and synced to your MP3 player without you lifting a finger. You can wake up each morning with new shows on your MP3 player ready to listen on your way to work.

This ease and convenience of this automatic delivery is powerful. This is what sets podcasting apart and made it so popular. Imagine walking into Starbucks to get a drink and by the time you leave you have new shows to listen to on your MP3 player. We haven't reached that point yet, but we are headed there.

Podcasts Automatically Put New Listening Content on Your MP3 Player on a Regular Basis Without You Having to Lift Even a Finger

At the core a podcast is an audio file that is automatically received from the internet and then synced to your MP3 player. The files are received by subscribing to what's called a podcast feed.

Sometimes you also hear this called an RSS feed, but that is just geek speak that gets used to try and impress you.

Maybe you have seen the small orange icons that look like this or this on some web pages.

This icon is a link to a news feed. Blogs use this kind of feed to deliver new content to their subscribed readers.

The same kind of feed is used to deliver podcast shows to your computer.

You don't need to know how the feed works to subscribe. You just need to know that you can subscribe to a podcast feed to get the latest shows when they are released.

You don't have to remember to check for new shows. You don't have to download it. It's done for you.

You can think of a podcast as being like a magazine because you subscribe to it and receive new content regularly.

Even though subscribing to a feed is the most convenient way to receive podcast, you don't have to subscribe to listen.

Many podcasts can also be streamed from the web or downloaded directly as an MP3 file.

Start Your Own Podcast.

Since podcasts showed up in fall 2004 thousands of shows have been created.

In the summer of 2005 iTunes added features for podcasts. This generated a huge leap forward in the number of podcast listeners and podcasters.

A popular podcast feed management site, Feedburner, reported that the number of listeners in January 2005 (13,500) almost doubled to 24,000 by February of 2005. This stat is from just one slice of the podcasting feeds on the internet.

Podcasting has become a big buzz word on the web and in the media. New services for listening to and creating podcasts are constantly popping up.

This industry is seeing very exciting growth. Podcasting is moving forward at the speed of light.

Despite this rapid growth, we are only scratching the surface. You can still catch this wave. Think of how fun it will be to create your own podcast and become a celebrity on the internet.

People will be listening to what you have to say. They will be enjoying your own creation. You can become and a recognized expert on the topic of your choice. This makes podcasting valuable for business use.

You can leverage this relationship with your audience to make money and promote yourself.

Your listeners will become dedicated and will return often for your opinion and expertise.

A Forum for Your Business

Podcasting allows businesses to share information with key audiences and, in some cases, to establish themselves as thought leaders in a particular area or on a particular topic.

The downside of this benefit, though, is that there are a lot of podcasts out there for people to choose from; you will need to work hard to find a specific niche that will allow you to stand out from the crowd.

If you search the iTunes store, for instance, you will find literally thousands of podcasts on every topic imaginable.

7 Benefits of Podcasting for Your Business

Podcasting can be an important addition to your business because you can:

1. Leverage your knowledge and content in another medium
2. Reach a different group of people (some prefer audio, some written word, and some video)
3. Offer more value to your current leads, customers and followers
4. Allow access to material at a time and place that suits your listeners (not you). They can download it and listen while exercising, on their way to work...
5. Be more engaging than with the written word (especially if you aren't a good writer) because the voice can express emotion more clearly.
6. Increase trust, and develop your relationship with your customers/followers – by hearing your voice they feel they know you better and that you're being more open and honest with them.
7. Establish yourself as the expert in your niche/field help you achieve market leadership

Here's the short description of the podcast process:

- Record sounds
- Transfer to computer (if necessary)
- Edit sound, adding voiceovers and other elements
- Compress into MP3 or other format
- Post to a Web server for distribution

Rather than reinvent the wheel, I recommend that you visit the site: ***The Beginner's Guide to Podcast Creation*** at http://tinyurl.com/4822pg

It's got a great tutorial and step-by-step instructions.

Another location to get information education plus a free location to broadcast your podcast from is Podcast Alley www.podcastalley.com.

They have resources for audio content creation and a podcast directory.

Distribution (or Feeds) for Podcasting

The podcasting process is simple and requires that you upload your edited audio to a site so that it's readily available for downloading by visitors when they decide they want it.

As a podcaster, you create and update your RSS feed for your listeners to subscribe to.

An RSS feed is nothing more than a specially formatted text file. RSS stands for Really Simple Syndication.

This is just a set of rules that outlines how the information in the feed fill needs to be formatted.

You don't need to worry about understanding RSS or writing out the feed file by hand. There are ways to create it automatically. You won't be writing code or trying to remember the RSS format. This will be done for you.

An RSS feed for a podcast includes information about your MP3 file such as its Web address and file size.

The podcatcher uses this information to find and download the audio file.

Each time you post a new podcast episode, you need to update your RSS feed with information about it so your listeners' podcatchers will detect and download the file.

In order to share your podcast with the world, you'll need to post your video and an RSS file on the Internet.

However, you can't create your RSS file until you know the URLs for your video files, so you'll need to find a host, upload the video, then create and upload an RSS file.

You have several options for hosting your podcast.

- If you already have your own website hosted on a server that allows you to upload multimedia files, use that site as your host.
- You can host your files on a site that offers free storage for media files, such as Weebly.com.
- If you choose this option, you'll also need to go through the process of publishing your podcast yourself.

- Finally, you can sign up with a site that both hosts and publishes podcasts. Some popular sites include:

    - www.buzzsprout.com - Audio only - free 2 hours upload per month - 90 day life - paid version is $12 to $24 per month - no expiration
    - www.hipcast.com - free to try - audio and video podcasting - Paid version is $8 to $20 per month - $20 per month is 3 gig upload and storage and unlimited bandwidth (# of downloads) ***(Seems to be the BEST deal!)***
    - www.podcastrevolution.com - free to try - audio and video podcasting - Paid version is $5 to $25 per month - $15 per month is 1 gig upload and storage per month and unlimited bandwidth (# of downloads)
    - www.libsyn.com - audio and video podcasting - Paid version is $5 to $20 per month - $15 per month is 250 meg storage per month and unlimited bandwidth (# of downloads)

Each of these sites has simplified instructions to follow to host your podcast on one of each of the above options.

Now all you need to do is get your podcast noticed.

Hopefully, you'll begin to attach a growing audience immediately. Continue to update your podcast regularly with new episodes.

Also, check out other people's podcasts to stay current in the podcasting community. Networking can be the key to success.

Podcatchers for you and your listeners

A podcatcher, or podcast client, is a computer program used to download various media via an RSS or XML feed.

For all practical purposes they are a software program that combines the capabilities of programming the podcast that you want to record, software for downloading and storing it automatically, and an audio player so you can listen to it.

While podcatchers are most known for downloading podcasts (generally audio files in MP3 format), many are also capable of downloading video, newsfeeds, text, and pictures.

Some podcatchers can also automate the transfer of received audio files to a portable media player.

Although many include a directory of high-profile podcasts, they generally allow users to manually subscribe directly to a feed by providing the URL.

When Apple added podcatching to its iTunes software in June 2005, it almost immediately became the most popular client.

Most of the podcatchers are similar to the TiVo system you have on your television. It combines the ability to search for podcasts, program them for capture, capture them when they are broadcast, and play them when you want to listen to them.

The following is a link to some of the most noteworthy podcatchers or software with podcatching capability.

www.podcastdojo.com/podcast-managers-or-podcatcher

The article covers computer, IPhones, and Android apps so there's bound to be something there for every taste and need.

We strongly recommend that you recommend check out which one of these works best for your applications and recommend one or more of these to your clients so that they know where and how to get and install a free player for your podcasts.

Creating and maintaining your own dedicated web site

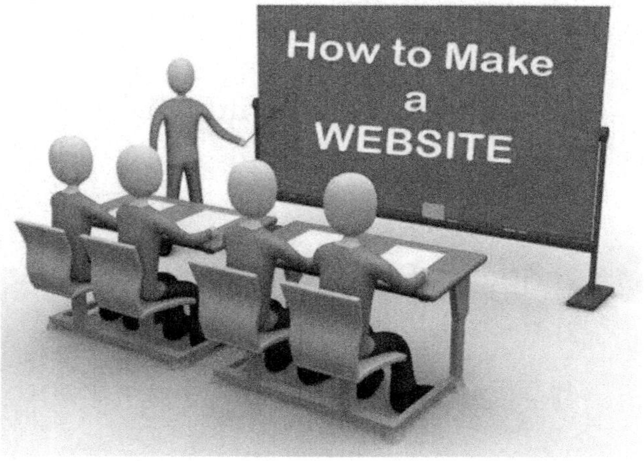

You've probably already noticed that both live audio and podcasting needs for you have your own web site and although you may already have a web site and be proficient in adding elements such as podcasting, I'm going to add this section to this manual for those of you who don't already have:

- A website
- A website that allows you to do all of the things you want to do
- A website that requires you to spend your life writing code or having someone do it for you
- A website that costs you a fortune and isn't doing what you want it to do.

Enter Weebly (www.Weebly.com)

Weebly is a web-hosting service that allows the user to "drag-and-drop" while using their website builder.

- Weebly offers a free online website creator. It uses a simplistic widget-based site builder that operates in the web browser.
- All of Weebly's site elements are drag-and-drop.
- Weebly works with Microsoft Windows, Mac OS and Linux computers and automatically generates a mobile version of each website.[16]
- Weebly lets site owners use a subdomain name if one is available. Users can buy a domain name from Weebly or re-direct it from elsewhere.
- The default site URL is "yourname.weebly.com" (yourname being whatever you want it to be as long as it isn't taken) but it can be changed if you already own a domain name.
- Weebly supports basic features for blogging and e-Commerce. Site owners can develop simple stores with payments through PayPal, Stripe or Authorize.net.
- Storage through Weebly is unlimited. However, the service restricts individual file sizes.
- Weebly users can choose to incorporate ads in their pages. Weebly supports visitor statistics tracking through its own tracking tool or Google Analytics.

Pricing Plans

In addition to offering users to build websites for free, Weebly has 3 premium plans including:

1) Starter Plan ($4.83 per month for 6 months),
2) Pro Plan ($9.83 per month for 6 months), and
3) Business Plan ($29.00 per month for 6 months).

All the plans are discounted even further if a user signs up for Weebly's 1 year or 2 years plans.

A user does not have to upgrade to a paid plan in order to build a very basic website using Weebly's drag and drop tools.

The benefit of subscribing to a paid plan is that more features are unlocked to a user. If a user wants access to more e-commerce building features, the Business Plan is most suitable.

To connect a custom domain name to a Weebly website, a user will have to at least upgrade to a Starter Plan.

Why use your own dedicated web site?

The short answer is that you can control what goes into it, how you run it, and because it's yours, you can keep all of the revenues you generate.

The long answer is because it's yours, you'll have the absolute control over what you want to offer and how you want to offer it – something that might not be available when you sign up for another service.

Adding Pay-per-download to your web site

For as long as the Web has been around, there has been a process called pay per download that allows the visitor to enter their credit card and download software, audio, video, and any form of electronically deliverable material

Recently a new service has been integrated into PayPal that provides this once very expensive service to virtually anybody who wants to use it. It's called E-Junkie and it's available from www.E-Junkie.com.

The philosophy is quite simple:

You register with E-Junkie

You fill out their electronic shopping cart

You upload the material to be sold via download

E-Junkie provides a link to the pay per download site so that you can put it on your Web page.

E-JUNKIE
we help you sell online

When the visitor wants to buy a downloadable item, such as audio, video, software, or a database, they simply click on the button and they will be linked to the E-Junkie site.

Once there, they will be required to enter their credit card information and once it's approved, they can download the material directly to their computer.

This is the same kind of system that iTunes and other related MP3 music systems have capitalized on so extensively in the last two years.

You can sell downloadable audio, video, seminars, PDF manuals, and virtually anything that you can convert to an electronic format.

The system has several different plans based solely on the amount of storage space you're going to use or the number of products you're going to offer.

There fees start at $5.00 per month and ramp up to $300 per month depending on the size or quantity.

The best part is that it's a fixed price whether you sell one or a million downloads each month so starting out with a free PayPal account and a $5.00 per month E-Junkie account will put you in the pay per download business for very little money.

Teleseminars and tele seminars with interaction

Teleseminars are used to provide information, training, or promote or sell products to group of people interested in a particular topic.

They are similar to traditional seminars, in content and purpose, but are given over a teleconference or bridge telephone line rather than at a specific location.

It is an emerging way to communicate, provide tele training, and conduct business without the cost of travel. The host of the teleseminar will schedule a specific time and date in advance to communicate with his/her audience.

The audience can vary in size from a few callers to 1,000 participants depending on the capacity of the bridge line used and the popularity of the topic being discussed.

These conference calls are typically recorded. There is typically a fixed period of time devoted to the presentation of information followed by another fixed period of time for questions and answers.

Teleseminars provide an opportunity for a host to provide information to a large number of people at one time. It allows a trainer to train many participants at once, one on many rather than one on one.

They eliminate need for travel, expensive preparation and presentation material costs which make teleseminars a very cost effective delivery method.

If the teleseminar is set up with interaction, the participants can be invited to ask questions or make commentaries all the while being monitored and controlled by the hosting party.

Teleseminars can be free or have a cost associated with participation for the students. Despite the participation fee, the advantage for students is this medium does not require the hassle and expense of traveling to a live seminar. Participants can join the teleconference from home or anywhere that there is a telephone connection.

After paying the fee, participants will receive a phone number and pass code for the call. If there is no charge for the teleseminar, the phone number and pass code may be distributed via email or may be available on the company's website.

Think of it as an old fashioned party-line telephone call where everyone who listens in gets charged.

What you can do with teleseminars:

- Introduce new books/products
- Get known in a new market — and get to know that market in turn
- Explain how a product or service works
- Interview experts and share their knowledge with your audience
- Answer questions about your area of expertise

In addition, teleseminars can help you do the following:

Sell more products and services. By presenting a presentation over the phone, you offer much more interactivity and a stronger personal connection than you can through a website, guest blog post or direct mail piece.

Create products. The calls themselves can be turned into products or classes. You can sell access to your live events, or you can record them and sell them later as CDs or audio downloads. And that's just for starters.

Get to know your audience. Being able to interact with your audience in real time allows an unprecedented level of market research, right in the moment.

Establish yourself. If you interview experts in your area, you will quickly be seen as an expert by association.

Build your list. Teleseminars are popular ways to introduce yourself to a new market. It's easy to invite people to your free events — and as they sign up, they become part of your list.

Create trust. Trust is crucial when you want to do business online. And one of the best ways to build trust with your audience is to interact with them directly. As they hear your voice live on the call — and as they ask questions and get answers right then and there — you become "real" to them and they'll be able to bond with you. This is something that's much harder to accomplish through sales letters and emails.

Free or fee?

There are about a gazillion services tele seminar available and I think I looked at almost all the free teleseminar tools.

Before we continue, remember that nothing is really free and these services are no exception – somebody is paying the freight – whether it's you paying for the services or the callers being charged through their phone provider, or all the participants having to sign up for an e-mail list and being solicited after the session..

So when we refer to free, we're talking about you not paying directly for the services.

There are several different services with different plans available so what we recommend is that you check out one or two of their web sites and make note of the services they offer and the fees they charge (and the way that they charge them) and weigh them against your own needs.

After that, you'll be able to make an informed decision as to which one is appropriate for your specific needs.

Free Services

Free Conferencing (www.freeconferencing.com) has been my favorite FREE tool. It has a lot of features like an online web controls, ability for Q&A, you can broadcast previously recorded audio files (great for guests with limited time slots available), you can use their optional screencast feature, you can break up participants into "discussion room", free recordings you can download or make available for playback, chat function plus, you can have up to 1000 participants and a 6 hour maximum time per call.

If you are interested in teleseminar services definitely check it out

Fee Paid Services

For paid services available, I recommend Instant Teleseminar (http://instantteleseminar.com/ProductInfo/?x=2670784

They give you a 21 day $1 trial to try out their services. This is pretty much a done-for-you service and is highly recommended.

Other options

Here are some others that are very good so check them out before you make a final; decision:

www.freeconferncecall.com
http://www.freeconferencing.com
www.unlimited**conferencing**.com
www.**conferencing**.us

Live Video

How to insert a video Feed onto a Web Page

Creating a stream from your webcam is a great way to connect visually with your clients and prospects.

This is probably best used when you are covering a live event, like a panel discussion or are giving a demonstration or presentation that's best viewed in real time.

The simplest way to create a stream from your webcam and embed it in your web page is to use a service for broadcasting live video, and then embed that video into your web page using your service's embedding tools.

After reviewing the available options, we concluded that Livestream, Ustream and Justin.TV are all well-established streaming services that allow in-browser webcam broadcast -- meaning that they work on any system with an Internet connection.

www.LiveStream.com
http://www.ustream.tv
http://www.justin.tv

You'll need a webcam, microphone, computer, and an internet/web connection that has enough speed to accommodate sending live video without looking jerky or jumpy.

The higher the upload speed is the higher will be the video quality you can push. So if you think about HD make sure you have at least 1MB upload speed.

High-speed internet choices

1. Cable Internet Speed:
Down speed (more is better): 25 to 100 megabits per second
Up speed (more is better): 2 to 8 Mbps
Cable internet is the fastest overall choice in 2011.
The best choice for broadcasting your own streaming video.

Depending on your location, you can get blazing fast download speeds of 30 to 100 megabits-per-second (Mbps).

2. ADSL: Asynchronous Digital Subscriber Line (aka DSL from the phone company) Speed:

- Down speed: 1.5 to 15 Mbps
- Up speed: 128 kbps to 1.0 Mbps
- Very convenient for users who already have telephone subscriptions.

ADSL achieves speeds that are not as fast as cable, but can be quite fast for most users: 8 to 15 megabits per second

Once you have your equipment set up, the basic ground rules are simple:

You'll need a camera and microphone that interfaces with your computer. Webcams, including the ones in your laptop computer, are already configured to do this but you can use a camcorder and separate microphone too if you want better quality. There's a great article at https://ustream.zendesk.com/entries/21550068-Basics-of-streaming-video-production that will give you insights into what works best.

You'll need a fast computer (which ones aren't these days?☺) and a high speed internet connection.

You'll need a space to shoot your video (See the section on creating great video in this manuals)

You'll need to sign up for an on-line broadcast service.

You're on the air!

Instructions for Broadcasting a Stream

Choose a streaming service that best meets your needs and comfort level, and create a new account. All three services referenced offer free accounts with advertising on the website and occasionally during the stream; you can remove the ads by purchasing a paid account.

On Justin.TV, click "Create Account" at the top of the page to fill in your account details.

On Ustream and Livestream, click "Sign Up."

Create your channel. Your channel is a place where visitors can view your feed in real time on the streaming service website, and where your old videos will be archived.

On Livestream, once you're logged in click "Channels" and select "Launch a New Channel."

For Justin.TV, you do not need to create a new channel -- you are your channel.

On Ustream, click "Create a Channel" on your account dashboard.

Begin your stream. All three services will require you authorize your webcam and microphone before use.

On Justin.TV and Ustream click "Go Live" in the upper right corner. Justin.TV's interface starts once you allow the service. Ustream will first ask what channel you want to use; select it. Ustream's Web broadcasting will pop-up in a new window. When you're ready to go live, click "Broadcast Now."

In Livestream, click "Start Broadcast" on the upper right corner. Like Ustream, the Livestream Webcaster pops up in a new window. Click "Go Live" when you're ready to start your stream.

Embedding Your Streaming Video. Find the embedding code for your channel's video.

In Justin.TV, click the "Share" button below the video on your channel page and copy the contents of the "Embed" box.

For Ustream, go to your Dashboard and click "Channels" > "Advanced." Click "Copy Embed" for the first box to only embed the video; click it on the second box to embed the video and chat together.

In Livestream, click your username to get to the account center. Next to the channel you want to embed, click the "Embed" button. Choose the widget you want to embed on your page, as well as the layout and the size. Copy the embed code when you're ready.

Open the page you want to embed your webcam stream on. If you're creating a new page for your stream, open a fresh text editor and design your page accordingly. If you're using a design program like Weebly™, simply add a button and insert the code for an instant link or simply paste the embed code where you want it to appear on the page.

Save your page and upload it to your web space.

You can now share your live feed page via email, social networking sites, or any conventional method that shows the recipient where to view you streaming video.

If you're streaming at a specific time, don't forget to add that to the notification.

Camera and microphone recommendations:

Since you're operating the entire process through your computer, you'll need a camera and microphone system that interfaces properly with your computer.

Although you can use a regular webcam that sits on top of your monitor, or the built in camera and microphone that's embedded in your laptop, if you're looking for quality and versatility, you'll be a lot better off using a regular camcorder and microphone.

In the section of equipment recommendations, we've referenced several models that we have tested if you're interested in acquiring new equipment.

On the other hand, you can spend $35 bucks and get an audio/video adapter box that will allow you to use your existing camcorder and microphone as a USB input device.

ION Video 2 PC Analog To Digital USB Video Converter for PC ($35 bucks)

www.amazon.com/ION-Video-Analog-Digital-Converter/dp/B002MT7JFQ/ref=sr_sp-btf_title_1_10?ie=UTF8&qid=1405544385&sr=8-10&keywords=usb+audio+adapter+for+pc

Video e-mails

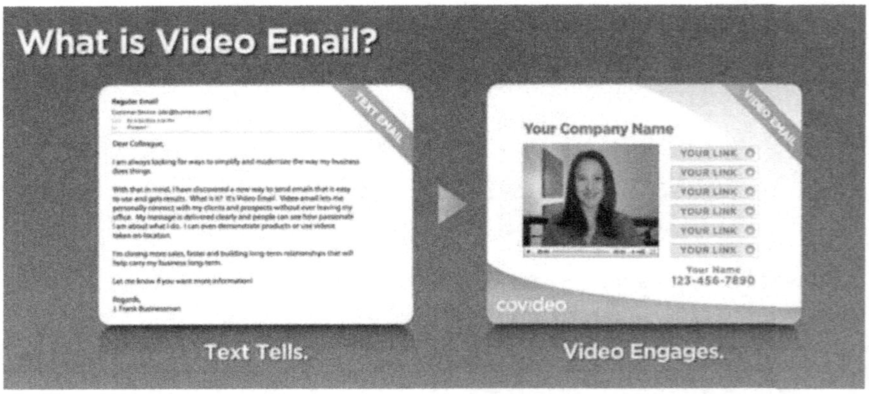

Video email is a messaging system where a webcam is used to record a video message which can be sent as an email.

Some video mail services require a download Client (computing) to work for the sender, while others offer a web-based option that allows the sender to send a video email without a downloaded client.

Video email is much like text messaging except that the sender can include a video.

A video messaging system provides access to the recorded message through one of its designated service providers.

The user then can retrieve the message through the same service provider, and access this message from anywhere in the world.

The sender records a message using their webcam or other technology, and sends to the receiver who can then access and respond to the message.

Messages can range from personal to business related, and can also be sent to either one or many receivers.

Depending on video message service provider, an account may or may not be required to send and receive messages.

Sending video email requires two fundamental steps:

Creating a video file

Attaching that video file to an email.

Creating a video file requires a camera and microphone as a minimum, with recording software as a possible extra. Most new computers come pre-installed with all requirements and many computers are video mail ready.

Sending video email allows for the recipient to have an easier time understanding the sender.

Although they can't respond in real time, in many cases they are able to send another video email back in response.

Video email eliminates the need for typing and composition, which may be difficult for some people. It eliminates the requirement of proofreading, and concerns about how a text email may be interpreted.

People can use tone of voice and physical expressions to help a message come across more clearly.

As technology continues to advance, it has gotten easier to get started with video email. Most programs do not require a download, and many of the services are still free to the public.

All that's usually needed is an email account, a registration process and a webcam to record outgoing videos.

Many of the services are free to try and free to use if you have a small number of recipients.

If you want to pre-record your video or send to a large database, you'll be getting into the paid services.

How to Select a Video Email Service Provider

The following is a list of 15 factors to consider when selecting a video email service provider:

1. Ease of use - The benefits of video mail is meaningless if the service is not easy to use. Complicated systems and processes, downloads, and multiple steps will place a service in the 'unused' category. You should be able to quickly see how easy the service is to use from the service provider's home page without having to sign up.

2. Video Email Open and Click-through Rate - If your video emails are not played, you have wasted your time and money. Ask the service provider for their video mail delivery rate, open rate, and click-through rate. The best video email providers are great at email delivery, keeping off of ISP black lists, and complying with the CAN-SPAM laws.

3. Video Quality - All service providers should allow you to try their video email for free, without signing in, so you can judge quality. Is the image clear? Does the video stream smoothly or does it buffer? Is there a delay prior to playing? How does it look on a computer, versus a smartphone or tablet?

4. Volume of video emails - Some service providers provide unlimited video sending and playing, others limit you to a pre-set number of video mails before they charge a higher amount. And other providers will charge based on bandwidth, which is impossible for you as a user to measure. Consider your needs on frequency of use, and your desire for a fixed cost each month.

5. Video storage capacity - Does the provider limit video storage, and if so is it by megabytes or number of videos? It can be confusing to the user to attempt to add up the size of all their videos just to stay within a limit. Also determine if the provider counts a 500Mb video upload as 1 video, or if it consumes your entire storage allotment.

6. Video Mail Branding - Some service providers have lots of holiday and occasion templates to use, which is fun in the short term, but the novelty quickly wears off. Cluttered graphics will detract from the video message so beware of services with complex, colorful graphics that overwhelm the video. For business use a look for a video email provider who allows you to add your branding in a professional look. A simple uncluttered look with your branding can add value to the video without being distracting.

7. Mobile Playback – Many people are accessing email on their smartphones. Do the service provider's videos play smoothly and clearly on mobile devices? Can your provider record on an iPad and have it play back properly on an Android, and vice-versa? Be sure to test before making a selection.

8. Mobile Video Recording - Smartphones and tablets have excellent cameras, making it easy to record and send video email on the go. Does the service provider have mobile apps, and if so for which platforms? Is the app easy to use? Do mobile videos get saved in an online account? Can you place your branding on mobile video emails?

9. Video Sharing - In addition to video email you may want to embed videos into your website, use them in your CRM, place them in Facebook and LinkedIn, Tweet videos, upload them to YouTube, or simply download them to your hard drive. Consider if the provider you are evaluating has these options, and in what service plans. Consider also if the provider includes a link that you can place in a 'regular' email; this can be quite useful.

10. Useful features - Beyond the basics of getting your video email delivered, consider other useful features such being notified in real time when someone opens your video. What about having the ability to control the number of views or number of days before a video email self-destructs? It can also be useful to add links to your social media sites in your video email template. Consider the degree of control and privacy. With mail VU when you delete a video from your account no one can play the video any more, even those to whom you've already sent it. This ability allows you to retract video email instantly. Other services don't give you this control, allowing recipients to continue playing videos no matter what you do. If you ever said 'oops' after sending an email, you will really appreciate the ability to prevent viewing!

11. Video Expertise - Consider if the video email provider is an expert at video recording, encoding, hosting, and streaming, or if they use someone else's platform. Does the provider host videos on their own service, or do they host their videos on YouTube? If the provider does not host videos on their own platform it is a warning sign and you need to ask why. Does the service provider offer other video services such as video hosting, video contests, or video testimonial recording, and do they have a Video API for developers?

12. Support - Consider your personal needs for support and help on things like Email delivery, or technical support for common webcam troubles. Does the provider have email and/or phone support, or is it completely self-help with FAQ databases? When you call the company do you speak with an engineer?

13. Type of company - Some video email service providers are multi-level marketers, others are marketing companies, and others are engineering companies. Consider how the provider makes money. Is it selling service subscriptions, is it advertising, or is it signing up sales representatives? Google each company being considered.

14. References - You should be able to find other users of the video email service provider. Just Google the provider and look through the links. Is the provider listed on The Top 101 Most Useful Websites on the Internet? Have they won any other awards? Read through the provider's blog and see what topics are being discussed.

15. Pricing - A single sale made as a result of sending video email can easily exceed the cost difference between any of the providers. Select the service that give you the greatest likelihood of achieving the results you want, rather than trying to save $5 here or there. A small price difference won't matter if the video quality is poor, the service hard to use, or the video emails are not viewed.

Now that you have the criteria, go evaluate the video email service provider field and decide who can best serve your needs.

No matter whom you choose, video mail will provide the emotion and personal connection missing in text emails.

Here are some suggested providers for you to try. Many have free trials:

www.streamsend.com
www.covideo.com
www.eyejot.com
www.videoemail.com
www.iwowwe.com

Video Podcasting

Although Apple had nothing to do with the podcasting technology, the iPod and podcasting quickly became linked and now with the versatility of video on cell phones, the iPad, and a host of other devices, the world of video just expanded by orders of magnitude.

The major advantage of a video podcast is the demographics. Podcasting enables people to reach millions of viewers outside who have the chance to view your programs when they choose.

Moreover, listeners have the privilege to again watch a program that has already aired besides being able to play or pause and to rewind or fast forward.

While audio podcasts can be a great tool, video podcasts allow you to add a visual dimension to your material.

Creating a video podcast is a surprisingly simple process. You can get started with not much more than a computer and camcorder Video

Podcasting is basically a way of publishing and distributing video files on the Internet using RSS.

Podcasting in simple words is an RSS feed which contains a video file.

Users can subscribe to podcasts using aggregators or readers, which periodically check for updates and downloads new content. It can then synchronize the content to the user's portable device like a iPad or tablet computer.

However, podcasting does not necessarily require an iPod.

Any digital video player or computer with the appropriate software iTunes can play podcasts which includes Smart Phones and tablets.

In other words, podcasting gives users the freedom to choose what they want to listen and when.

This brand new way of broadcasting content is still misunderstood by many publishers, even though it's growing rapidly.

So, what is a video podcast?

A video podcast is a streaming or downloadable video clip, usually offered as a subscription to an ongoing program.

Like audio podcasts, video podcasts can be on any subject and last any length of time.

They can be dramatic, comic, fact-based, or news related. Some will offer video walkthroughs of how to use certain products or how to make something.

They can be a great enhancement to technology or gadget review podcasts, as they allow the viewer to actually see the products and get a better understanding of how they work.

If you would like to make your own video podcast, you will need some basic equipment.

In addition to a digital video camera, a microphone, and a computer, you must have a high-speed Internet connection and the ability to create an RSS feed, which is the broadcasting instrument.

After filming your video podcast you can give it a more professional feel by using video editing software.

There are several online guides to making and broadcasting your video podcast to help you through the basic process.

Creating a Video Podcast for distribution

The short course is as follows:

Shoot the video
Edit the video
Convert the video format for distribution
Distribute the video
Making a Video Podcast

This video provides an overview of creating a video podcast, including how the process of uploading it online in the most efficient, cost-effective way.

http://www.youtube.com/watch?feature=player_embedded&v=h91TL_-5GQw

The video discusses the two ways of recording video with a video camera, editing the video, exporting the final product into iPod format, and finally the importance of creating a podcast blog for storing the video in order to avoid the large bandwidth fees charged when you post to a regular web post.

The video also highlights several websites that are great for hosting, and then walks you through the process of putting your blog on iTunes.
Digital Video Cameras

Now that we've covered the details on getting your video on-line for people to view, we need to backtrack a bit and talk about the hardware you're going to need to get your video material into your computer to start with.

You can use the video recording feature on your smart phone, a webcam for your computer or a separate camcorder.

You can get a decent digital camcorder on EBay.com for $40 bucks and for most general applications, it'll work just fine.

For the serious videographer, there's an entire spectrum of camcorders available including digital still cameras with a video mode. Some of the big DSLR cameras are currently being used by professional studios instead of conventional video cameras so there's lots of good equipment at reasonable prices.

I'm going to make a strong suggestion in when it comes to the audio and zoom features.

Get a camera or camcorder that has an optical, and NOT a digital zoom
With an optical zoom, you are really filling the screen with close up and the definition and clarity improves as you zoo in.
With a digital zoom, you're really taking a wide angle shot and then doing a mock zoom by spreading out the pixels which results in a fuzzy picture why you zoom in.
Get a camera that has an external microphone connection jack and get a good microphone to use when you record
The microphone on the camera will pick up the sound around the camera which is probably not where the subject or the action is taking place so you need to have a microphone that's located where the camera is looking.
If you use the one that's on the camera, you'll get all of the background noise between the cameras and the subject which can destroy the overall content of the recording.

Check out the resource section for our recommendations on cameras for podcasting and virtual presentations in general.

Posting your video podcast

Rather than trying to be tech savvy, it's lot more expedient to simply sign up with a site that both hosts and publishes video podcasts. Some popular sites that we can recommend are:

www.hipcast.com - free to try - audio and video podcasting - Paid version is $8 to $20 per month - $20 per month is 3 gig upload and storage and unlimited bandwidth (# of downloads) ***(Seems to be the BEST deal!)***

www.podcastrevolution.com - free to try - audio and video podcasting - Paid version is $5 to $25 per month - $15 per month is 1 gig upload and storage per month and unlimited bandwidth (# of downloads)

www.libsyn.com - audio and video podcasting - Paid version is $5 to $20 per month - $15 per month is 250 meg storage per month and unlimited bandwidth (# of downloads)

Each of these sites has simplified instructions to follow to host your podcast on one of each of the above options.
Webinars

What is a webinar?

The basic concept of a Webinar is that instead of you standing in front of an audience lecturing and answering questions, you appear in front of a video camera and the audience is in some remote location interacting with you via their computer.

Contrary to what you might have heard, the correct technical definition of a webinar is: *a live online educational presentation during which participating viewers can submit questions and comments.*

This means that when you conduct a webinar, it's supposed to give the virtual audience the ability to:
See you and/or your materials
Listen to your materials
Ask questions about the materials and
Leave comments about the materials.

These elements differentiate the webinar from other forms of video presentation formats because they are ***supposed to*** include interactivity options that allow the audience to actively participate in the program.

The operative words here are ***supposed to*** because many of the webinars that are offered do NOT activate the interactive elements of the system because they don't have the time or the interest in allowing the audience to participate.

Nevertheless, the format allows for interactivity whether you choose to activate it or not.
How to Choose a Webinar Provider
In addition to your computer gear, you'll need a service provider to host and distribute your webinar and choosing the webinar provider that's right for you and your clients depends on what you want from the service and how you wish to pay.

Some services offer a payment-by-participant, while others offer monthly service fees and contracts.

Review your clientele and purchase the webinar service based upon the clients who will use the system to access the webinar. If the majority of your clients are not technologically savvy, then you want to find a user-friendly webinar provider. On the other hand, if conducting a webinar for people who are more comfortable with technology, then a more advanced system would be acceptable.

Calculate the budget for the event. To re-coup your webinar costs, ensure the fee covers the cost of the webinar, your time and allows room for profit.

Repeated monthly webinars may warrant a service that provides a monthly rate. If you plan to conduct one or two webinars per month, this might be a better option for your program.

Become familiar with the software and webinar requirements in order to host a glitch-free event. There's nothing worse than hosting an event that no one can access.

Check the free trials from a variety of webinar providers and practice using their services. Some providers offer freely weekly or monthly trials. Match your webinar needs with the provider for a better overall experience.

Select a company that includes a replay feature in the webinar you offer your clients for those who pay for the webinar but may not be able to make it on the scheduled date. This allows clients to download the webinar on their schedule and will result in much more participation.

Verify the provider can handle the amount of users planned for your webinar. Many webinar services crash when offering live streaming access and too many users try to log on. Ask providers for the contacts of other customers that could talk about their experiences with the company.

Recommended webinar providers: Rather than try all of the available services, we turned to http://webinar-services-review.toptenreviews.com for their current list of recommendations for the best providers. Here are the five we've tested and we suggest that you check out their website for full details and comparisons.
Go To Webinar www.gotomeeting.com
Webex www.webex,com
Onstream www.onstreammedia.com/webinars.php
Fuze www.fuze.com
Adobe Connect www.adobe.com/products/adobeconnect.html

Steps for Planning a Successful Webinar

Conducting a successful webinar involves lot more than just turning on a webcam so here are some of the major steps you can take to plan quality, affordable webinars:

1. Ask yourself (and others) whether a webinar is the right tool for your needs.

Webinar software can be a powerful training and outreach tool, but, like all technologies, the decision to use it should be subordinate to your strategic objectives and the needs of your audience.

While webinars work well for some topics, they're not suited to every training need.

When determining whether a webinar is the best medium for your needs, consider your audience, the subject matter, and the time you'll need to cover your topic.

If you're addressing a small, internal audience of colleagues about a new organizational goal, for example, a webinar may be a less appropriate option than it would be for, say, a training seminar for a large audience of clients and funders.

Likewise, a daylong course on effective interpersonal communication might not translate well to an online seminar.

Most online audiences tend to lose interest after about two hours, so organize your topics and information so they can adequately be covered given this time constraint, or break your curriculum into chunks of two hours or less.

When you're breaking a large topic into smaller chunks, leave at least 15 to 30 minutes between sections.

However, if your audience is busy, or if they're unaccustomed to online learning, it's usually best to leave a day or more between sections.

In other words, few of us can spare more than two hours a day for training and even that is difficult to manage.

When evaluating whether a webinar meets your needs, you may also wish to solicit the feedback of subject-matter experts and webinar tool experts, other organizations in your field that have conducted their own webinars, and even the audience you plan to address.

Informal conversations, formal interviews, and surveys and polls can all help you assess whether a webinar is the right medium for your — and your audience's — needs.

2. Recruit speakers and a support team.

Once you've decided that a webinar is the right tool for you, you'll need to assemble a team of staff members or volunteers to help you run it.

Whenever you see a TV show or a Movie where there is a talk radio show host, you also see one or two other persons in the control booth and for many webinars, that's how you're going to look.

You probably won't have a glass encased booth with a big mix down board and lot of equipment, but you will need someone to handle the entire button pushing logistics that are involved with conducting a successful webinar because for most people, it's just too much juggling to do it alone.

In general, there are three main players in a webinar:

the organizer/facilitator,
the presenter or presenters, and
Assistants.

You might fill all three of these roles yourself, you might assign them to three different people, or you might need more than three.

Bear in mind, though, that while some presenters can handle all of these roles on their own, we only recommend the going-solo approach for an experienced trainer addressing a small audience.

In most cases, you should at the very least divide up the organizer and presenter roles, and for large, complex webinars you'll often need one or more assistants.

Organizer/facilitator. The organizer is the person responsible for developing the webinar topic, locating a speaker, marketing the event, setting up the registration, and communicating with participants.

The organizer usually participates in the webinar itself by introducing speakers, interviewing the subject matter experts, moderating audience questions, and encouraging audience participation. The organizer also monitors the overall situation and helps to troubleshoot logistical and technical problems. In other words, the organizer does most of the hard work, and most of the steps outlined in this article. Time commitment: roughly 10 to 20 hours per webinar.

Presenter(s) (also known as subject matter experts). Ideally, presenters should concentrate their efforts on preparing their demonstration and their PowerPoint slides.

During the webinar, their main focus should be the presentation itself, as well as fielding questions from the audience. Worrying about the webinar software, event registration, troubleshooting, and other logistical details detracts from the presenters' ability to give an engaging presentation.

In certain formats (a panel discussion, for example), you might have more than one presenter. Time commitment: four to six hours per hour of webinar.

Assistants. Experienced organizers often produce webinars without any assistance, but there are at least three scenarios where you should consider asking for help:

When you or your audience is unfamiliar with webinars and webinar tools;

when you plan to play a large role in the conversation (either as an interviewer or participant); and
When you expect a large audience.

Assistants can help by answering questions that the organizer and the presenter don't have time for. Often, assistants focus entirely on answering technical and logistical questions *("I can't hear the audio," for example).*

In this case, you can recruit and train the assistant yourself, and anyone moderately tech-savvy can do a good job.

However, for webinars where you expect a large audience and where you anticipate more audience questions than the presenter alone can answer in the time allotted, you may need an assistant who understands the subject at hand to help with answering chat questions.

An alternate approach would be to refer all unanswered questions to an online forum where the presenter can respond at a more leisurely pace. Time commitment: one to two hours per hour of webinar.

3. Determine the format.

A webinar can be structured in a variety of formats to suit different purposes. Below are some popular formats you might consider:

Type	Description	Pros	Cons
One Speaker	A single presenter speaks, demonstrates, and answers questions from the audience.	Fewer people to coordinate and train on the Webinar tool.	A lone presenter is more likely to become the authority at the "front of the room," which might make some in the audience reluctant to participate and ask questions.
Interview Style	Interviewer asks a set of predetermined questions.	More engaging to hear multiple voices. The fact that the interviewer is asking questions of the expert(s) often encourages the audience to do the same	More people to train and coordinate. Scheduling the run-through and the actual webinar may be more difficult.
Moderated Panel Discussion	Multiple people on the line at the same time, with a moderator facilitating the discussion.	Offers a variety of voices and perspectives.	More people to train and coordinate. Scheduling the run-through and the actual webinar may be more difficult. Can be challenging to keep panelists from talking over each other
Interactive	Audience members participate fully via instructor-led exercises and facilitated conversations.	If done well, participants receive a deeper understanding of the topic because they're fully engaged in the dialog and the exercises.	Can only accommodate a small group. Requires a very skilled, experienced teacher/facilitator.

Plan your visuals.

Since webinars rely on audio and visuals to get the message across, both should be engaging.

Plain slides with a lot of text don't work as well as interesting visuals that illustrate the topic being discussed.

Examples of visuals you may wish to include with your presentation are:

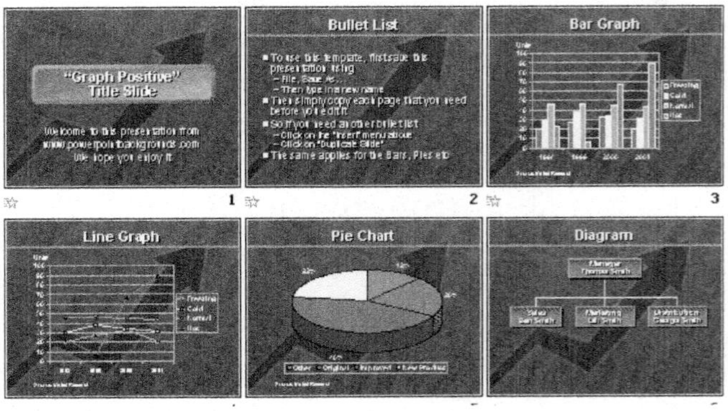

- An introductory slide reminding your audience how to log in to the audio and what time the webinar will begin.
- A slide introducing each presenter, including job title, affiliation, and a photograph if available.
- A quick overview of the webinar agenda and the topics to be covered.
- Screen grabs of websites or tools you will be discussing. If possible, try to show the sites and tools in action (rather than just the image stills) for a more dynamic experience.

Note: Most webinar tools allow you to share your desktop, displaying in real time your interaction with programs and websites.

5. Pick a tool.

There are dozens of web conferencing tools out there, offering a variety of features tailored to different needs.

Check out the systems we've outlined above and give yourself some time to try out various web conferencing platforms.

Once you get started with a particular webinar tool, it's hard to switch to a new tool.

There are several reasons for this:
You and your colleagues have learned to use the software,
Your regular participants are familiar with this tool, and
You've already paid for a subscription to this program (unless you're using a free tool).

When weighing your software options, here are a few questions to ask:

How many people will the tool accommodate?

Most tools and pricing plans set a cap on the number of participants. Once you reach that number, new participants find themselves locked out or the tool lets them in and charges you a fee for each person above the cap.

For free and low-cost web conferencing packages, the cap is often as low as 15 or 20 participants.

Other plans top out at 50 or 100, while enterprise-level packages allow as many as 1,000 participants per webinar.

How much does it cost?

While some packages are free, other vendors charge for web conferencing and audio separately, some charge per participant per minute, and others charge a flat fee per month or per year.

When you add in the fees for hosting recorded webinars and the cost of a toll-free phone bridge (when applicable), the pricing schemes can be complicated.

Which features will you need?

Do you want to just show a presentation or demonstrate how to use a specific piece of software?
Would you like your participants to be able to take control of your desktop?
Do you want a live video feed of the speakers?

Make sure you choose a tool that allows you to do what you want.

How is audio handled?

Some products offer integrated, web-streaming audio, which allows participants to listen to the presentation through their computer speakers or their computer headsets.

With this arrangement, if participants plan to talk, they'll need a microphone for their computer or a headset with a built-in microphone.

If they work in close proximity to their colleagues, they will also need computer headphones or a computer headset to avoid bothering their neighbors.

A headset with microphone that plugs in via USB or standard audio ports is well-suited to this type of webinar and costs between $20 and $40.

Other webinar platforms require that participants and presenters dial in to a special phone number, often referred to as a phone bridge.

You usually have two options for this:
a toll number, in which case the participants have to pay the fees charged by their long-distance or cell phone provider; and

a toll-free number, in which case you or your organization will pay five to ten cents per minute for each participant.

If you absorb the costs of these calls, be aware that they can add up quickly during webinars with a high turnout.

For example, 53 people participating in a 60-minute webinar at six cents per minute would cost a total of $191.

Finally, many webinar platforms offer you both web-streaming audio and phone-bridge options.

You can choose to enable one or the other, or both. Enabling both types of audio gives you and your audience some flexibility in the event of a technical problem.

Do you want to record the webinars and make them available for later viewing?

YES! YES! YES!

In my experience, one of the major problems with the virtual world is NOT making the material available on-demand when the user has the time and incentive to use it!

I'll grant you that not having to drive is a tremendous cost and time advantage but if your programming is scheduled when there is a conflict with the listener's schedule, you've got a problem.

By recording the material and making it available on-demand, you eliminate this problem.

Be sure to ask how the software handles recording and whether the vendor charges extra to make that recording available on the Web.

Most vendors charge for the Web hosting, rather than the recording feature itself, but you should always check to be certain.

Monthly hosting fees can be as high as $15 per month per recorded hour.

If you feel the fees are excessive, you can download the recording and make it available yourself, but serving or streaming the recording from inside your network might put a strain on your bandwidth or your technical expertise.

When evaluating software, you will also want to ask what exactly gets recorded.

Some tools, for example, only show the slides along with audio, but don't record the chat conversation or the desktop sharing.

Vendors also vary in terms of how long they save the recording. Some delete it after a month, while others save it until you delete it yourself.

Create an agenda.

About a week or so before your webinar, hold a conference call with the speaker or speakers and determine what questions you'll ask and the order in which the speakers will present.

If you are using a format that enables visuals, ask each speaker to furnish the graphics and images to accompany his or her section of the presentation well ahead of time. (If speakers are demonstrating software, only a few slides will be necessary, as most of the webinar will likely be spent on the application itself.)

You may also want to ask speakers to send a photograph and brief biographical description that you can use for registration and outreach.

Follow up this initial call with an email containing notes from your discussion. These notes may include:

A list of topics and questions you came up with during the conference call.
Deadlines for materials. If your presenter plans to use PowerPoint slides or other visual aids, ask them to send you the graphics and visuals at least two or three days before the run-through so you have time to proofread and merge the materials with your own.
For example, you may have a PowerPoint template you use for all presentations and webinars. You may have introduction and conclusion slides you want to add. And, of course, you should double-check each slide for typos.

An agenda with the order of the speakers and the duration of each segment

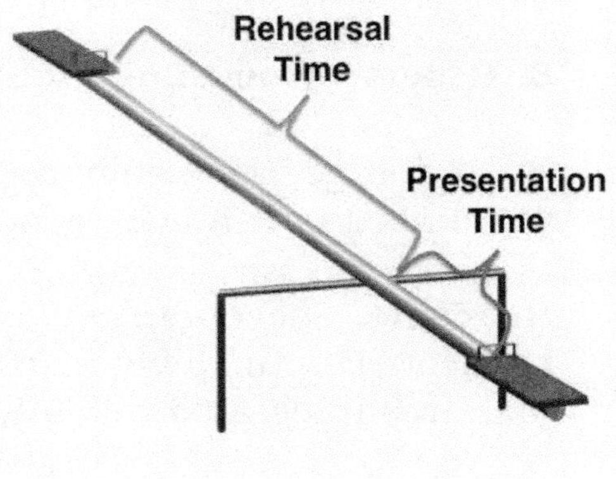

Schedule a dry run.

If possible, sometime before your webinar, you should schedule at least one run-through with all participants to work out any unresolved questions or technical issues.

Your dry run should cover the following:

An introduction to the participants. If speakers haven't met during the initial call, this would be a good time to introduce the people who will be present during the webinar, both online and behind the scenes. Make sure the participants know whom they should turn to if they have questions during the webinar — and how they can reach them.
An introduction to the webinar tool and its features. Discuss how to use the tool, what features are available to the presenters, and where participants can go to get extra practice in using the tool on their own before the event.
An equipment check. This is a good time to ensure that all of your presenters' operating systems, web browsers, headsets, and other equipment are compatible with the web conferencing tool.

Fortunately, most webinar tools let presenters and participants test their computer for compatibility before the event itself.

You can usually do that by directing them to a web page where they can launch a wizard that tests for the necessary components and plug-ins.

Be sure to include instructions for this with your registration information.

While you can't force attendees to check their computers, follow up with your presenter to ensure that they have all of the downloads and plug-ins they need.

A review of your agenda and visuals. Go over the agenda and move through the presentation to ensure that slides are in the right order and that speakers know where they come in.

Before the run-through, compile your presenters' visuals and load them into the conferencing tool.

This will help presenters understand what the attendees will see, as the presenters' view is different from the attendees' view.

A dry run is also a great opportunity to generate enthusiasm for the upcoming event and rally your presenters.

8. Reserve your equipment and space.

By and large, the webinar tool you choose will determine the type of equipment you'll need to run it. In general, you will want to have the following:

Headsets. The organizer and all presenters will need headsets — telephone headsets if your webinar tool uses a phone bridge or computer headsets if your tool uses integrated web audio.

Technically, you can use the telephone handset if your webinar package has a phone bridge, but it's distracting and tiring to lift a phone to your ear for an hour or more, especially when you're using a mouse and keyboard.

Never present a webinar using a speakerphone. Even in a quiet location the audio quality is often poor, and in noisier spots a speakerphone will pick up background noise.

A power adapter if you're using a laptop, as a backup for your battery.

A network cable to connect you directly to the network if you aren't using a wireless connection.

Regardless of the equipment you use, you will need a quiet space in which to conduct your webinar.

In addition to using a headset, you should reserve a conference room or place where there won't be background noise or interruptions.

In addition to keeping out background office noise, you'll also want a space secluded from outside distractions, such as construction noise or sirens.

9. Set up a system for registering attendees and determine your price structure.

Before you begin marketing your webinar, determine what tool you will use to register attendees.

While most fee-based webinars are offered in the $25 to $40 range, others can be priced at upwards of $200 per attendee.

Keep in mind that while organizations that invest significant time and energy into a regular series of high-quality webinars might recoup some of their costs by charging attendees, nonprofits that only produce a few webinars a year may drive away potential participants with a fee, not to mention creating extra work for themselves for very little profit.

On the other hand, charging a modest fee ($5 to $20) to recoup the cost of your time and expenses might be acceptable, depending on your audience.

As you assess your audience's needs (see Step 1), you can determine whether they are likely to pay for the type of webinars you'll be offering.

If you plan to charge a fee that your audience will see as significant (for example, more than $25), make sure most or all of the following are true:

You have a wealth of hands-on experience, an especially deep knowledge of the subject, or some other attribute that makes your advice especially valuable to potential participants.

You have a lot of experience with training in general and online communication in particular. *Keep in mind that participants have higher expectations when they pay; the more money you charge, the greater the expectation that you will deliver an engaging, well-produced webinar.*

You have the capacity to collect payments easily using a tool like PayPal.

You plan to conduct webinars on a regular basis.

One advantage of charging for your webinar is that it provides an incentive for participants to show up.

If everyone pays beforehand, you'll have fewer no-shows. With free webinars, you can expect that roughly 50 percent of the people who sign up will fail to attend.

Part 7 Support Materials

Creating Handouts for Your Audience

It's doubtful that all of your virtual presentations will be standalone programs. On the other hand, it's more likely that they will require some form of support materials such as note, manuals, or workbooks.

If you are giving a presentation including handouts or workshop materials, it will benefit you to take the time to prepare and print those materials properly.

It may take the form of a packet of printed out slides, or it may include worksheets, sales literature, or other matter. The printed part is what your audience will take home and remember, so make sure that you get them looking good.

What to include in the handouts

Rule #1 - The Presentation Handout is NOT your slides

The handout is an amazing thing. It frees your mind, it (can) improve your slides, it makes your audience happy! The audience will be happy; because they don't have to write stuff down while you are talking (true multitasking is a myth, in case you don't know it yet).

They can pay full attention to what you're saying. And should they have questions a week after the presentation, your contact details are on the handout.

The handout – what should it look like?

So, how should the handout look like? It should be preferably 1 page long (up to 3 pages is still ok, but shorter is better, otherwise people won't bother reading) and it should describe what you have talked about in the presentation.

It can also include graphics if you deem it relevant; maybe a key image from the presentation slides is a good idea – people will then remember your presentation when looking at that image.

What NOT to do: It is currently quite popular to hand out the slides as a handout, causing a symptom of what many people call the "slideument" –

an attempt to kill two birds with one stone. Even if you include your notes in the slides printout, they're still quite ineffective. So avoid the slideument.

What goes in the handouts?

How many times have you referred to handouts that you received at a presentation, only to find that they didn't contain enough detail for you to recall what the speaker said?

This is a typical experience when the handouts consist only of pictures of Microsoft Office PowerPoint slides with bullet points.

For a handout to be useful when the presentation is but a distant memory, the handout must contain enough detail for it to stand on its own.

Let that sink in for a minute: ***The handout should stand on its own.***

The purpose of the handouts, then, is to provide the audience with the additional materials that they need to support and act on the key points of your verbal presentation.

Delivery Methods for handouts and support materials

Paper

With virtual programs, it's doubtful that you will be supplying the handouts in any type of physical format although you'll want to consider that the participants will want a printed copy so format them for the standard 8 ½" x 11" paper that will most likely have in their printers..

These can be as simple as a single page outline through a complete commercially available manual, workbook, or text.

On-Demand Download

The current technology available on web sites has made it practical to upload and store your support material and then allow the participants to download and utilize it as necessary.

The material can be stored in both hidden and password protected locations that require a password if the materials are sensitive.

Since the material is being delivered electronically, it can contain multi-media, audio, video, and PowerPoint slides plus, you can provide the materials in full color at no extra cost.

Large File transfer Systems E.G. – Dropbox

If delivering large volumes of material from your website is not currently a practical or viable option, you can use a third-party service that will allow you to use the electronic equivalent of a bus locker to store your material and then send the recipient a link to it so they can download the material for themselves.

Having used the services many times, I can tell you that they are extremely convenient and take the trepidation out of offering to share something that you created with somebody else especially if it's something proprietary.

I'm going to recommend that you use the ones that I've outlined below because they'll let you do a certain amount of store-and-forward work for free. Then, if you find that this is really valuable to you, it'll be worth your while to spend a few dollars each month to sign up for their fee services.

Look closely at your projected usage and actual costs. The pricing plans are not always as obvious as they seem and you might end up buying a lot more capabilities than you need.

- www.Pando.com
- www.Sendthisfile.com
- www.yousendit.com
- www.dropbox.com
- www.dropsend.com

NOTE: On any materials that you are distributing electronically you must consider that both the Widows and Macintosh software will be used by the recipients.

Consequently you MUST deliver the material in a universal format such as PDF or HTML if you expect it to be read and utilized by both platforms.

Part 8 - Cloning and Distributing Your Program

The world of virtual training and communications has three parts:

1. Creating the material
2. Distributing the material and
3. Promoting the programs.

In the forgoing sections, we've explained the different ways you have available to create your own materials so now we need to look at how you're going to get the materials you've created out to the public.

Regardless of which of the methods you choose for your virtual programs, it all boils down to one of two ways to deliver the materials:

1. Live or Recorded and
2. Physical or virtual.

This may appear to be a bit simplistic but it's actually a bit more complicated than it appears especially if all you've ever heard about or been involved in are live presentations.

Once you decide on which technology set you're going to use, you'll need to decide how you're going to make the material available to the public.

Here are some of the considerations that we'll explain so that you can decide which ones are best for your particular needs.

- Real time presentations
- On-Demand Presentations
- Free programs
- Fee Paid (Pay-Per Download)
- CDs, DVDs and Flash drives
- Commercial Seminar Services

Let's take a look at these different ways of distributing your material so that you can pick and choose which ones are suitable for which type and more importantly, what will work best for you.

	Type of Materials to be Delivered			
Type of Distribution	**Text**	**Graphics**	**Audio**	**Video**
Real time presentations	Yes	Yes	Yes	Yes
On-Demand Presentations	Yes	Yes	Yes	Yes
Free Programs	Yes	Yes	Yes	Yes
Fee Paid (Pay-Per Download)	Yes	Yes	Yes	Yes
CDs, DVDs and Flash drives	Yes	Yes	Yes	Yes
Commercial Seminar Delivery Services	No	No	Yes	Yes

This matrix is based on a combination of fifty years of experience in the music business, video tape sales, infomercials, and professional speaking/training.

You'll also notice that with the exception of the commercial seminar delivery services that you can use just about anything in the way of materials you want.

When you get into the commercial venues, simply offering text and graphics are not recommended because what you're doing is basically offering the old mail order teaching courses using the latest in technology.

I've seen several programs of this type and every one of them has been greeted with rejection and people have demanded refunds.

If you want to sell an e-book, e-course, or e-manual use one of the other delivery methods.

Now let's look at the pros and cons of each of the six delivery methods method as they currently exist.

Real time presentations (Virtual Delivery)

Imagine yourself giving a live speech or a lecture in a classroom where all eyes are on you right now.

Now imagine that there are cameras and microphones connected to web and that everything you say and do is going out live to the audience.

That's a real time presentation.

If you are a seasoned professional speaker and know your material like the back of your hand, this can be a powerful tool for you and your business enterprises.

There are those extemporaneous moments where things just seem to work out right and your material takes on a new sparkle and level of delivery that's so good you wonder if it was really you doing it.

On the other hand, you don't get any *'do-over's'* and what you do and say goes out to the audience complete with any errors or technical problems you encounter.

As a professional speaker/trainer who does over 150 live presentations each year, I can attest to the fact that doing it live is always more exciting but it isn't always the best.

Things happen, things get forgotten and left out, and sometimes, the program is simply a disaster for reasons that are out of my control.

I've done a lot of live programs that were recorded for broadcast or resale as downloadable programs or for CD/DVD distribution. Some were really great and some were not.

On the other hand, whenever I get the chance to do a pre-recorded program where I can work with a script, props, and know that if I screw up, I can do it over, the presentation always seems to turn out better.

On Demand Presentations (Virtual)

These are programs that can either be live or pre-recorded and are recorded and stored and offered for viewing and/or listening by the audience at a time that's convenient for them.

Obviously, recordings of live presentations are going to be subject to the same concerns as if they are going out one time but have the added concern about their material being stored and consumed forever.
A lot of people think that the best way to address this process is to ONLY do pre-recorded and edited materials.

Don't sit there and think that just because you can redo something that it's going to be easier because it's not.

I've seen more stage fright involved in recording sessions than I care to remember.

I started my career in the studio music business and saw great musicians choke during a recording session because they felt they were under pressure to perform perfectly.

As a professional speaker, I saw speakers and trainers who had done the same material dozens of times blow it because they were too focused on doing the material perfectly and would become stiff and forget critical components.

The 'do-over' part helps a lot because if you can calm down after doing a couple of takes, you'll find that the material actually gets better due to the rehearsal time that you've just given it.

If you've ever seen the outtakes from TV shows and movies, you've seen some of the funniest moments ever recorded but what you don't see is the endless hours of error correction and frustration involved in getting it perfect.

If you've ever been on a professional video, TV, or movie shoot, you know that the directors demand retakes until the talent's (that's us) eyeballs bleed and our tongues turn to cotton.

The big difference between these shoots and what you're most likely going to be doing is that these professional shoots involve a lot of money and they can't afford to have it be anything less than perfect.

For example, on my last infomercial shoot for Bon Clayman (The man who did the Ginsu Knives), we were pitching a book and CD set but every time I held up the CD, it was off center.

After about a dozen takes, we ended up using masking tape to hold the CD in the palm of my hand so that when I raised my hand it was perfectly aligned.

Sounds like fun, doesn't it?

I'm not trying to persuade you one way or the other but I do recommend that you look closely at your skills, nerves, and ability for doing it perfectly the first time.

Free Presentations (Virtual or physical)

This one is a little weird because you might think that it doesn't belong in a delivery vehicle category but quite frankly, that's where most people thing about it so that's where we've put it.

In addition, a lot of people think that free or fee doesn't' make any difference – the material still has to be good. Or does it?

This is one of those areas where the opinions fly freely and get very controversial because one hand you'll hear: *"It's free, who cares how good it is?"*

The other school of thought says: *"It doesn't matter, if the material is bad, even if it's free, bad is bad"* or *"That was worth exactly what I paid for it, nothing!"*

Sometimes doing a quick and dirty recording or live webcam program is critical for the specific needs of the audience and it won't seem to matter if you stutter, forget something, or act disjointed because all they care about is getting the material.

On the other hand, what sucks, sucks, and people remember your performance regardless of whether they paid cash for it on not because they invested their time to review your materials.

Have you ever heard the statement: *"Well, there's an hour of my life that I'll never get back"?*

There is NO qualification as to whether it was free or fee – it was still bad and your reputation suffers accordingly!

YouTube® is loaded with recordings of instructions on how to do a lot of things especially when it comes to using high tech things like software and related computer stuff.

Some of it is well done and a pleasure to watch and some of it is pure junk where one viewing will keep visitors from ever using the material created by that individual again.

The moral here is that free or fee, if you're are serious about doing virtual programs, I admonish you to look carefully at your skills and if you are not the kind of presenter that can get it right the first time or are good enough to "fake" your way through a bad presentation, do a recorded program where you get "do-over's" until you get it right.

Pay-Per-Download (Virtual)

These are pay-as-you-go courses in everything from basic math skills, to computer programming, project management, and other skill sets that people have a need or interest in attending but can't get away to a formal classroom.

For example, *Ed 2 Go*™ is an on-line seminar company that has its programs listed in most of the college catalogs in the United States.

Since the material is based on an academic model, they get away with the presentations being mediocre since most professors are not the greatest speakers or most dynamic orators.

This is the opposite of a free presentation and MUST be perceived as worth the price the user is paying for it.

In the case of the on-line courses, most of the material is a combination of text and graphics (similar to a workbook or manual) but more and more, the material is staring to include narrated PowerPoint™ presentations and in some cases, full video productions.

Nevertheless, the material is good enough to fill the needs of the audience and as long as it covers the subject adequately and doesn't put the audience to sleep, it's accepted and remains a part of the course offering catalog.

Because many of these are for course credit, all of these programs are pre-recorded and theoretically edited and corrected before being offered for sale.

I'm sure that some of them are first time recordings but I also suspect that the bulk of them have been scripted and edited to get rid of mistakes, pauses, and other objectionable elements that will make the audience uncomfortable or cause them to stop viewing.

If possible, you'll always be better off if you offer rehearsed programming whenever you are charging for it so that you can edit it for content, delivery and completeness.

You can add this feature to your own website though a pay-per-download service like E-Junkie or the built-in digital download features in Weebly's e-commerce shopping cart.

This type of program delivery allows you to deliver virtually anything that you can convert to digital format.

CDs, DVDs and Flash drives (Physical)

Companies like Nightingale-Conant and others have sold tens of millions of dollars' worth of audio cassettes, Video Tapes, CDs, DVDS, and Flash drives.

Couple that with all of speakers, trainers, and training companies that have offered their wisdom, experience, and insights for a fee and you have a ten billion dollar business.

Because much of this material is used for reinforcement of a given topic, it lends itself to audio, video, text and graphics plus other things like software, legal forms, data bases, templates, and instruction manuals.

Over the years many platform professionals have sold recordings of their live performances and done very well with them. Some of the performances were great, once in a lifetime moments, that were captured for posterity and some of the material was pretty bland.

Nevertheless, it was of interest to the buying public and once made available for purchase, sold a lot of copies.

We started with a six set of audio cassette tapes plus a manual in what's known as an album and later on we converted everything to CDs because the technology changed.

Today, CD players are available for less than $20 bucks and most vehicles come with CD players built in, so this is a great way to distribute your material for mobile listening.

Of course, video material has to be viewed in a static location so DVDs are perfect for this type of program although recently, some of the material has been put on flash memory drives and sent through the mail.

If you're serious about delivering your material for sale in a 'canned' format that will likely be around for a while, I recommend that you do pre-recorded and edited material whenever possible.

Of interest is that of the dozens of programs that my partner and I offer, only a couple of them are from live recordings – everything else is from studio recordings that we've outlined, scripted, performed and edited before making them available for sale.

Commercial Seminar Delivery Services

If I said that this whole manual was directed towards this specific area, I would be telling you the truth because if you're serious about getting your material delivered without a lot of hassles, this is the best way to do it.

My years of owning an international mail order marketing company taught me a lot about inventory costs, inventory control, cash flow, accounting, and a whole host of other business skills that are invaluable in my current business enterprises.

Let's start with the three parts to a successful business enterprise:

Getting the business
Doing the business and
Running the business.

As the person who is creating and delivering the material, you're the one who's doing the business.

Once you've got the material *"in the can"*, you'll want to spend your time getting the business, the part that includes advertising, promotion, and sales aspects which is where the money is generated.

What you do NOT want to be doing is handling the day to day elements of running the business other than to monitor your cash flow and spending.

This is where the commercial seminar delivery services excel because once you create your program(s) and store the material with them, everything else, except for the advertising and promotion is automatic and handled by them.

Here's the short course on how they work:

1. You sign up with them and open an account
2. You create your program or materials
3. You upload the materials to their web site
4. You promote the programs that are stored on their web site

When a client places an order on their web site, the delivery service charges the client's credit card, delivers the material via the web, notifies you of the transaction, and makes the payment available to you for electronic funds transfer.

By offloading the mechanics of the process to an outside firm, you have now freed up your time to either promote your programs or to create new materials.

Before we go any farther, I'm going to get on my soapbox and tell you that many of the firms who offer these services are a waste of time and closely resemble a scam or rip off.

Having been one of the Internet/Web gurus since the 1990's, I have been hustled by dozens of firms that claimed that they were going to be my financial salvation because they were offering me the opportunity to deliver my programs through their service – for a huge fee, of course.

This is simply not true for a variety of reasons the most dominant being that as the number of offerings grow, your program will be lost in the masses and it's doubtful that they're going to put a big push on your materials.

The better way to go is to sign up and list your materials with a company that charges a percentage for each download and do the promotion yourself.

Most of them no longer exist and of the ones who do, most have never produced enough revenues for the presenter to offset the costs of the services they provide.

In the resources section, I have listed a number of programs that we have investigated and the one that we are currently using and recommending.

The ones that we do NOT list or recommend are the ones who promise that you'll become wealthy through their promotion efforts.

Another recommendation is that whatever system you think you want to utilize has a *'try before buy'* policy.

If you're like most of us, you'll be dazzled by their ads and promotional campaigns where they show you how quick, easy, and effective their systems are.

In reality, most of them are overly complicated, require a lot of technical expertise, and tend to become more of a part of the problem t=rather than a part of the solution.

This complexity causes most people to abandon the system and end up paying for something that you don't use.

Since you're planning on investing some money and a lot of time into this method of distribution, you'll want to be comfortable with the technology that you decide to utilize.

For example, I edit my video using a software package call Pinnacle Studio 12 which is a few years old and by todays' standards, is not the latest cutting edge state of the art technology. (They're up to version 17!)

On the other hand, I know how to use it and it does everything I need a video editing system to do including green screen and special effects.

I have looked and sadly, purchased more complex video editing software but frankly, I don't want to invest the time to learn how to be a professional editor and prefer to stay with the old system because it works well for me and my needs.

So now let's take a moment and review the four generic technology packages and see which one is going to be the best ones for you to use.

- Text
- Graphics
- Audio
- Video

Now that we've reviewed the tools that are available to us, let's look at the three (3) generic areas where the world of virtual communications works best.

1. **To maintain** *communications*
2. **To** *learn something* **new or different**
3. **To** *teach something* **new or different**

Here's a statement of fact: Virtual presentations are most often used for teaching and training.

Now before you proclaim that you're planning on using it for executive communications, and not for training, look at the definition of training:

Training is the acquisition of knowledge, skills, and competencies as a result of the teaching of vocational or practical skills and knowledge that relate to specific useful competencies.

Training has specific goals of improving one's capability, capacity, productivity and performance.

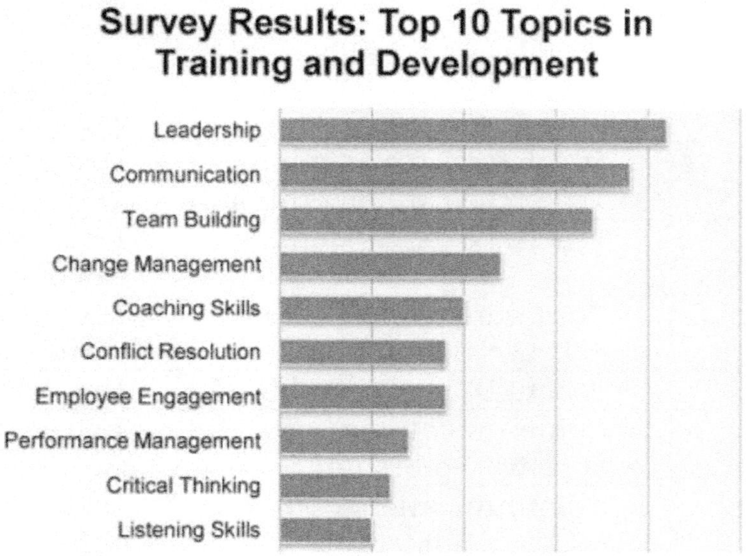

Training also covers the need to continue training beyond initial qualifications: to maintain, upgrade and update skills throughout working life. People within many professions and occupations may refer to this sort of training as professional development.

The bottom line here is that if you're communicating with someone else, you're basically training each other about the topic being discussed, so the term training is all encompassing rather than limiting.

Here is a sample list of generic training and communications topics with the accompanying method of virtual content delivery that is currently deemed as the most effective technique and technology.

Topic	Primary Area of Interest	Best Delivery Method(s)
Business	Communications	Text, graphics, audio, video
Design	Teach Something	Text, graphics, audio,
Development	Teach/Learn Something	Text, graphics, audio,
Health & Fitness	Teach/Learn Something	audio, video
Humanities	Teach/Learn Something	Text, graphics,
IT & Software	Teach/Learn Something	Text, graphics,
Language	Teach/Learn Something	Audio
Lifestyle	Teach/Learn Something	Video
Marketing	Teach/Learn Something	Text, graphics, audio, video
Math & Science	Teach/Learn Something	Text, graphics
Music	Teach/Learn Something	Audio and Video
Office Productivity	Teach/Learn Something	Text, graphics, audio, video
Personal Development	Communications, Teach & Learn Something	Text, graphics, audio, video
Photography	Teach/Learn Something	Text, graphics, video
Social Science	Communications, Teach & Learn Something	Text, graphics, audio, video
Teacher Training	Communications, Teach & Learn Something	Text, graphics, audio, video
Test Preparation	Communications, Teach & Learn Something	Text, graphics, video

Although the list is generic, it shows two significant things:

1. That the bulk of the virtual communications applications are for training and development and

2. That even though this list is generic that just about every topic you can think of, including yours, whatever it may be, is a potential candidate for virtual training and communications.

This generic list is taken from the list of currently available seminars from Udemy (www.Udemy.com).

If you visit their site, which is one of our highly recommended resources, you can click on any of the genic topics and a drop down menu will appear showing you a variety of sib-topics that are currency being offered.

This extensive list should give you an even greater understanding of what's being offered and should spurt your own creativity in deciding what you want to create and how you want to deliver it to your audience.

Last, but certainly not least, as you review the resources section, be mindful of what it's going to require for you to create and offer your materials in a virtual format.

Do NOT be seduced by the sexiness of methods like *'green screen video'* and the latest in '3D technologies' just because it's cool and you can play with Hollywood style special effects.

This form of delivery takes a lot of work and is a major undertaking.

On the other hand, if you can deliver (hopefully for profit) a report with text and graphics, it'll be a lot faster and easier for you start your virtual delivery services.

This is not to say that you shouldn't do green screen video, but that you'll want to conscientiously evaluate your needs, the needs of your audience, your skill set, your budget, and what the ultimate goal is before you commit to a particular type of delivery system.

Part 9 - Getting the message out

Virtual training sessions usually depends on sending out an e-mail blast to tell people about it and get them to attend.

The problem is that with the deluge of e-mail that we receive, we MUST find a way to stand out or our message won't get opened let alone read, so we started doing detailed research on what works and what doesn't.

If you're not aware of it, the subject line is probably the most important factor in getting the recipient to open and scan (and hopefully read) your message as opposed to deleting it immediately.

Although I wrote the book (literally) on writing adverting copy *(Advertising Copy for the Clueless®)*, I started doing additional research on specifically what works and what doesn't in today's e-mail deluged marketplace.

My research led me to this comprehensive report conducted by an Australian company that really puts it all together.

It gets a little detailed in parts but overall, it's really easy to understand and more importantly, to apply.

Pay careful attention to the type of message like Business to Business (B to B) vs. Business to Consumer (B to C) and how different subject headers are used to capture better results.

152 killer keywords for email subject lines (and 137 crappy ones)

This edited opening copy and the accompanying report are reproduced by permission from: https://econsultancy.com

The entire report is included in the appendix

Introduction:

We had a hunch that word choice in email subject lines have a strong effect on response rates. So, we tested 287 keywords across a sample of 2.2bn emails to see which work, and which don't.

Why? Because President Obama has done more for email marketing than any world leader in the history of mankind. How?

By focusing on subject line testing, his digital team optimized their donation campaigns to generate hundreds of millions of dollars online.

Despite Obama's best efforts, most marketers still view email marketing as the Bluth Company's Banana Stand of Arrested Development fame: a more boring and less sexy marketing channel than pretty much anything else imaginable.

But – and never forget this – there's always money in the banana stand! There is great power in optimizing subject lines.

We tested a random sample of 95,000 global, English-language campaigns over the last 12 months (for a total of 2.2bn emails), and have isolated 287 popular 'trigger words'.

Then, split by sector, we looked at the correlation between the word's inclusion in the subject line and its variance above or below the average results for key email metrics (Open Rate, CTR, CTOR, and Unsubscribe).

To ensure outliers aren't confusing things, we've also looked at the first, second and third quartiles to give an indication of data spread, not just nominal long-run means.

Bear in mind, these relationships are correlated, but not necessarily causal.

There are simply too many variables in an email campaign to pinpoint exact causation. But, where there's smoke, there's fire, and at the very least, this will give you many ideas for what to test.

And now some key findings:

People love free stuff

In related news, water is wet. Specific trigger words have a drastic effect on the response rates of offers. For example, 'Sale' delivers +23.2% opens, but 'Save' only +3.4%.

However, in terms of click through, they give +60.7% and -25.2% respectively.

When you promote your offers, consider testing those trigger words – it could make or break your campaigns.

And, consider simple quick wins like 'Free Delivery' or even '% off' – in the right context these keywords can drive massive response uplifts. But, if you do nothing but hard sell and offer discounts all the time, your customers will become bored and tune out.

Mix up your offer emails in a series of value-adding campaigns.

Content marketing works when the content isn't crap

The problem with content marketing is the vast majority of content produced is crap.

Too many people have outsourced it to agencies that don't know enough about their clients' markets, and focus on the wrong metrics. As a result, consumers have become anesthetized to content.

Take, for example, 'Report' (-23.7% opens, -54.8% CTR) and Webinar' (-16.6%, -70.7%.)

Conversely if the content is good, people will consume it. So 'News' (+34.8%, +47.7%), 'Bulletin' (+15.8%, +12.7%) and 'Video' (+18.5%, +64.8%) work well.

Simply put, if your content is crap, it won't work. If your content is good, you'll get great results. When doing content marketing, make sure you're in the latter group, not the former, or else you'll be what we affectionately call in the industry a "*spamming %$*!$#*".

More frequent emails are better than less frequent

Typical for an email guy to say, right?

Full disclosure, when people send out more emails, ESPs make more money, this is true. But, if emails drive response and therefore generate revenue, then what's the problem?

But, don't take my word for it. We looked at newsletter frequency, and specifically the trigger words that indicate it. 'Monthly' brought -26.6% opens and -37.0% clicks. 'Weekly' brought +27.1% and +50.6% (not bad) and, amazingly, 'Daily' brought +27.8% and +100.3%. Simply put, more email drives more response.

And it's not just us saying this, check this out if you're skeptical.

Personalization works, if your data is clean

Lots of people have played around with subject line personalization, with varied results. We found the average opens and clicks decrease (-20.7% opens). But, the spread of the data is massive. The first quartile is -73.1%, but the third quartile is +30.8%. Why is this?

The main thing about personalization is to ensure you don't deliver users a disjointed user experience. If the subject line is personalized, but the email content isn't, guess what? You may gain opens, but have done nothing to drive clicks. It leaves users with a negative experience.

Furthermore, if the subsequent online journey isn't equally personalized, once again you've left your users feeling like you've given them the old bait-n-switch.

Most importantly, if your database is old and potentially incorrect, you're in trouble. Eyeball your data first and pick recent, engaged data for testing personalized content.

There's always money in the banana stand

This report analyzed a vast sample of subject lines. More than anything, it showed there is lots of short-run variance in every keyword sample. What matters is that you test things out to your lists, and never stay standing still.

What works one week may not work the next, but if you aren't trying out new things then you're treading water and ultimately throwing away money.

Email still delivers the strongest revenue of any digital channel by a country mile (source). In all your tweeting, pinning, and Facebooking, are you making sure you dedicate enough time to your email subject lines?

Remember – of Obama's 30 person digital team, 24 of them were focused on email.

They knew that there's always money in the banana stand.

Part 10- References and resources

So now that you've reviewed the world of virtual training and communications and have decided that it's the perfect tool for your needs, you're going to need some additional resources to make the technology come alive for you.

Everything we've discussed and explained are systems and delivery venues that we've personally used over the past few years.

To assist you with your entry into the world of virtual programming, we've researched a variety of software, hardware, and resources for acquiring what you'll need and made some strong recommendations based on both our research and personal experience with them.

We've divided them up into the different types of systems, services, hardware, and software we think will probably work best for you as well as setting up some complete system shopping lists so that you'll know what you'll need for each type of delivery system.

We've included reference articles and web sites that relate to the world of virtual training plus resources for reasonably priced video cameras with external microphone jacks, green screen studio kits, and a variety of components to help you create great virtual programs.

Text *(In addition to one-on-one messaging, we're also considering that you'll be making a blast to a list so you'll need some form of system that allows you to designate multiple recipients for the same message)*

E-Mail	
Method	**Recommended Resource**
Plain text with and without attachments	**ACT** (www.act.com) **Constant Contact** www.ConstantContact.com
HTML -With embedded links and graphics including attachments of text, audio, video and data	**ACT** (www.act.com) **Constant Contact** www.ConstantContact.com
Video E-Mail for a single recipient Including both live and pre-recorded materials	***Eyejot (www.eyejot.com)***
Video E-Mail for multiple recipients Including both live and pre-recorded materials	***Eyejot (www.eyejot.com)***

Graphics *(In addition to simply delivering graphic materials we're assuming that you will be conducting some form of narrated graphic presentation)*

Graphics	
Method	**Recommended Resource**
PowerPoint	***PowerPoint*** ***www.microsoft.com***
Narrated PowerPoint	***Photostage*** ***www.nchsoftware.com/slideshow***
Photos and other graphics (Maps, charts, etc.) including Large files and high volumes of pre-recorded information	***Dropbox*** ***www.dropbox.com*** ***Transfer Bog Files*** ***www.transferbigfiles.com*** ***Media Fire*** ***http://www.mediafire.com***

Audio *(This method assumes that the telephone, rather than a computer, will be used for the teleseminars while the Audio Podcasts will be available through computers, tablets, and smartphones with web capabilities)*

Audio	
Method	**Recommended Resource**
Teleseminars – Live, both unidirectional broadcast style and interactive including recording of programming for later listening or in a unidirectional broadcast format	***Join Me*** ***www.join.me*** ***Free Conferencing*** ***www.freeconferencing.com*** ***Instant teleseminar*** ***http://instantteleseminar.com*** ***Free Conference Call*** ***www.freeconferncecall.com*** ***Unlimited Conferencing*** ***www.unlimitedconferencing.com*** ***Conferencing*** ***www.conferencing.us***
Audio podcasts – Both live and recorded	***BuzzSprouts*** ***www.buzzsprout.com***

Video *(This method assumes that the receiving party has audio and video capabilities on their computer plus a webcam and microphone for interactive participation)*

Video	
Method	**Recommended Resource**
Webinars – presented live in both unidirectional broadcast and interactive formats	***GoToMeeting*** *www.gotomeeting.com/* ***AnyMeeting*** *www.anymeeting.com/*
Webinars – Recorded and offered in unidirectional Broadcast format only	***GoToMeeting*** *www.gotomeeting.com/* ***AnyMeeting*** *www.anymeeting.com/*
Video Podcasts -- Both live and recorded	***Hipcast*** *www.hipcast.com* ***PodCast Revolution*** *www.podcastrevolution.com* ***LibSyn*** *www.libsyn.com*
Video Conferencing including one-on-one and multiple participants	***Anymeeting*** *www.anymeeting.com* ***Webex*** *www.webex.com* ***Go to Meeting*** *www.gotomeeting.com* ***Google Hangouts*** *https://tools.google.com/dlpage/hangoutplugin*
Video E-Mail for a single recipient Including both live and pre-recorded materials	***Eyejot (****www.eyejot.com****)***
Video E-Mail for multiple recipients Including both live and pre-recorded materials	***Eyejot (****www.eyejot.com****)***

On-demand seminar delivery services *(This method assumes that you will be creating your material and offering it for sale though a third party service)*

On Demand Seminars	
Method	**Recommended Resource**
	Udemy ***www.Udemy.com***

Additional resources

Product Name	**Source**
Audio recording and editing software	www.audacity.com
Video recording and editing software	**Pinnacle Studio** www.pinnaclesys.com **Camtasia** www.techsmith.com **Windows Movie Maker** http://tinyurl.com/av8bnzq **ILife** www.apple.com/ilife/imovie **Wax** www.debugmode.com/wax **Lightworks** www.techradar.com/downloads/lightworks
Inexpensive wired Microphones for podcasting ($20 to $50)	http://tinyurl.com/mlsqopo http://tinyurl.com/mcndquc http://tinyurl.com/nap6agg http://tinyurl.com/ptn245z
Best wired microphone for podcasting, interviews, and webinars ($89 to $128)	http://tinyurl.com/mm3g7qa
Wired collar clip microphone ($3 bucks)	http://tinyurl.com/kh35ps8
Headsets - $10-$40)	**Full ear cup style** http://tinyurl.com/lhsfu9z **Noise Cancelling Earbuds** http://tinyurl.com/mzrkrow
Inexpensive wireless microphones ($39 to $180) Nady, Azden or Samson	http://tinyurl.com/kdyej9u http://tinyurl.com/7xgj5vw http://tinyurl.com/kfq9l69

Reasonably priced video cameras with external microphone jacks ($389)	**Canon VIXIA HF R400 HD 53x Advanced Zoom Camcorder** http://tinyurl.com/lfuqomx
Free green screen editors	http://download.cnet.com/VSDC-Free-Video-Editor/3000-13631_4-75764187.html
Green Screen Tutorial	www.youtube.com/watch?v=luyOn5DdUZs
How to make a cheap green screen	www.youtube.com/watch?v=8pQXfvFEGto
Green screen systems	http://tinyurl.com/mrwp78b Check out www.ebay.com for green screen
FREE 'shoulder surfing' software to let you share screens	www.mikogo.com

PowerPoint - Articles and Tips

http://tinyurl.com/ojqsdgc

Podcasts - Articles and Tips

www.learningtimes.com/what-we-do/podcast-production/benefits/
yourbusiness.azcentral.com/podcasting-work-20403.html
www.howtopodcasttutorial.com/what-is-a-podcast.htm
www.brighthub.com/multimedia/audio/articles/77995.aspx

Tele-Seminars - Articles and Tips

www.aliciadunams.com/the-benefits-of-teleseminars/
http://tinyurl.com/nvvdz9d
www.trainingauthors.com/teleseminar-basics-for-authors/
http://thefutureofink.com/tools-for-teleseminar-hosts/
http://tinyurl.com/px62afg
http://tinyurl.com/oxd55qy

Webinars - Articles and Tips

http://blog.cathy-moore.com/2014/01/tips-for-webinars-or-virtual-training/

http://hsm.typepad.com/inspiringideas/2009/05/5-benefits-of-adding-webinars-to-your-learning-and-training-options.html
http://www.chrisg.com/reasons-webinars/
http://www.planninghelper.com/spv-231.aspx
http://www.ehow.com/list_7495125_advantages-webinar.html

Rounds Miller and Associates

6318 Ridgepath Court
Rancho Palos Verdes, CA 90275-3248
310.544.9502
www.RoundsMiller.com

Business, Technology and Organizing Training Specialists

PRODUCT CATALOG

Audio/Video Toolkit

Complete Windows® software series including: **Banner Maker** PDF with video instruction for making your own web site banner using Microsoft Publisher®; links and information for 100,000+ royalty free graphics, sources for free music for videos and slideshows. **CamStudio**™ Video Screen recording software. **VideoSoft**™ Complete audio/video editing and conversion suite including 55 separate conversion programs for video, audio, and mobile devices. **Audacity**™ Audio recording and editing software.

Audio/Data CD+ **$24.95**

Book Marketing for the Clueless®

Want to sell your books, CDs and DVDs for a profit? This audio/PDF CD includes databases of over 500 catalogs and outlets that market books and instructions on how to solicit your publications including how to be listed with Amazon.com for free.

Audio/Data CD+ ISBN 978-1-891440-49-6 **$24.95**
*eBook** ISBN 978-1-891440-94-6 **$9.95**

Business Training for the Clueless®

A complete course about becoming a six figure a year business trainer. Contains a complete PDF manual, workbook, training aids, exercise, games, on-screen timing software, and royalty free audio clips.

Book & Data CD+ ISBN 978-1-891440-85-4 **$69.95**
*eBook** ISBN 978-1-891440-96-0 **$9.95**

Cash From Your Clutter

A manual filled with resources to help you turn your excess stuff into cash. Includes information on what sells, where to sell your excess stuff, how to place a realistic sales value on your stuff, the best time to sell your stuff, how to donate for tax deductions, a special section on how to sell your timeshare property and much more.

*eBook** ISBN 978-1-891440-79-3 **$9.95**

The Clutter Bug Attacks Christmas Clutter™

If there's one time of the year our homes and lives become more cluttered, it's Christmas. Here are quick and easy ways to simplify Christmas, including how to: plan and organize your Christmas, get your card and gift lists under control, simplify your gift giving, handle the gifts you receive, and get rid of Christmas clutter!

Book ISBN 978-1-891440-74-8 **$15.95**
*eBook** ISBN 978-1-891440-84-7 **$9.95**

The Clutter Bug Attacks Junk Mail, Spam and Telemarketers™

Overwhelmed by Junk Mail? Spam? Telemarketers? These can all be stopped! Current, verified information with visual examples for quick and easy understanding. The first book ever to show you how to get off of political mailings, plus proven and structured processes for both eliminating and retaining communications.

*eBook** ISBN 978-1-891440-81-6 **$9.95**

***eBooks contain a complete book in PDF for use on all Mac & Windows PCs, including desktops, notebooks/laptops, netbooks & tablets.*
+All data files are in PDF format, playable/viewable on Windows and Mac, audio files are in Wave format playable on standard CD players.

Rev. 12-13

The Clutter Bug Investigates Coupons, Discounts, and Deals™

A complete manual describing the coupon process and how to use them to save 35 to 85 percent on your shopping. Includes complete detailed illustrations and examples on where to find coupons, types of coupons, store policies for coupons, deals and discounts, rebates and refunds, and rewards and loyalty cards.

Book ISBN 978-1-891440-77-9 ***$24.95***
*eBook** ISBN 978-1-891440-82-3 ***$9.95***

Clutterology® Getting Rid of Clutter and Getting Organized! 4th Edition

Revised and Expanded! A complete manual on how to get organized, set up and maintain manageable filing systems, and eliminate clutter that gets in your way. Provides some of the simplest, easiest and most practical advice on how to remove the clutter from your life and get organized.

Book ISBN 978-1-891440-62-5 ***$34.95***
*eBook** ISBN 978-1-891440-71-7 ***$9.95***

Highlights of Clutterology®

This audio CD has over 60 minutes of tips, tricks, insights, and stories about getting rid of your clutter and getting organized. It's ideal for reinforcement to remind you that getting organized is a step-by-step process that you can accomplish if you take it easy and stick with it.

Audio CD+ ISBN 978-1-891440-44-1 ***$19.95***

How to Become a Clutterologist™

Do label makers and shelf dividers make you smile? Use your aptitude for organization to change lives and turn your decluttering skills into a moneymaking career; become a professional organizer! Includes the tools and knowledge you need to succeed in the professional organizer industry: organizing specialties, understanding the Clutter-Hoarding Scale, how to structure your business for SUCCESS, business licensing and insurance.

Book ISBN 978-1-891440-56-4 ***$34.95***
*eBook** ISBN 978-1-891440-68-7 ***$9.95***

Consulting for the Clueless®

In today's business climate, consultants are flourishing while others can't find a job. This book will show you how to profitably leverage your experience into a six-figure income with minimal investment. Includes DVD with consulting seminar, PDF copy of Marketing the One-Person Business, and support software.

Book and Windows Software DVD ***$39.95***

Contracts and Agreements for Inventors

Over a dozen of the most utilized agreements to help ensure that what's yours stays yours. With the help of an attorney, they contain everything you'll need from a confidentiality agreement to work-for-hire agreements, assignment of rights, and partnership agreements. Comes with instructions for usage, filling out, and filing where applicable.

Data CD+ ***$19.95***

E-Commerce for the Clueless®

Complete manual plus instruction program on CD that includes Flash® video on what e-commerce is and isn't, what's required to get involved, how to set up a web site page for e-commerce, setting up a FREE PayPal account and product purchase buttons, and how to set up a pay-per-download account to sell e-books and other information products directly from your web site. Includes royalty free background music clips

Data CD+ ISBN 978-1-8911440-78-6 ***$24.95***

**eBooks contain a complete book in PDF for use on all Mac & Windows PCs, including desktops, notebooks/laptops, netbooks & tablets.*
+All data files are in PDF format, playable/viewable on Windows and Mac, audio files are in Wave format playable on standard CD players.

Fishin' With A Net, 8th Edition

Learn the elements of designing a Web site that actually works for you and can be created in less than four hours. Covers what the Web really is, what to put on your site to be successful, and how to link with the search engines quickly and easily.

Book ISBN 978-1-891440-97-7 ***$24.95***

Headline Creator™ Pro Suite

"Your headline can result in 80 percent or more of the effectiveness of your ad or sales page!" Automatically generates time-tested, proven, results-oriented headlines based on the greatest headlines in history. . . and does it in 17 seconds!

Windows Software CD+ ***$39.95***

How To Sell Your Inventions for Cash, 3rd Edition

Everything you need to know to be a successful inventor! Takes your idea from inception through the licensing process to a manufacturer for royalties. Learn how to protect your inventions using patents, trademarks, copyrights, and other legal instruments, determine if you're ready to offer your idea, and how to find and solicit manufacturers who are interested in your ideas.

Book ISBN 978-1-891440-59-5 ***$24.95***
*eBook** ISBN 978-1-891440-63-2 ***$9.95***
Audio CD+ ISBN 978-1-891440-28-1 ***$39.95***

Intellectual Property Protection for the Clueless®

CD contains 3 hours of audio plus 100s of pages in PDF format on trademarks, patents and copyright. Includes forms for filing without an attorney! *Bonus: How to Apply for an Innovation Research Grant!* This audio is in MP3 format playable with the Windows Media Player or comparable MP3 software.

MP3 Audio/Data CD+ ISBN 978-1-891440-67-0 ***$59.95***

Marketing the One-Person Business, 2nd Edition

A one-person business is different from any other because you have to do the business PLUS get the business. Contains complete information about setup, operation, independent contractor criteria and forms, fee setting, consulting, public speaking, seminars, contracts and agreements.

*eBook** ISBN 978-1-891440-88-5 ***$9.95***

Mechanics of Starting a Home-Based Business, 2nd Edition

A home-based business is a business whose primary office is in the owner's home. Explains the realities of starting and operating a home business and including resources for taxes, licenses, and advertising plus computer operated business you can start and run.

Book ISBN 978-1-891440-64-9 ***$34.95***
*eBook** ISBN 978-1-891440-70-0 ***$9.95***

Photo Manipulator Software

Shrink Pic™ software automatically reduces the size of photos for email, blogging and web galleries. No set up, no operating instructions, just send your photos normally and Shrink Pic™ does the work. Paint Shop Pro™ 4 is an easy to use photo editing program that let's you enhance your photos and create professional-looking images.

Windows Software CD+ ***$19.95***

Professional Speaking for the Clueless®

Do you want to be paid to speak? This book explains the REAL business of professional speaking and how to make six figures a year without huge marketing and advertising costs. Includes dozens of resources, databases, and complete explanations of how to locate speaking opportunities and market to them.

Audio/Data CD+ ISBN 978-1-891440-53-3 ***$24.95***
*eBook** ISBN 978-1-891440-91-5 ***$9.95***

**eBooks contain a complete book in PDF for use on all Mac & Windows PCs, including desktops, notebooks/laptops, netbooks & tablets.*
+All data files are in PDF format, playable/viewable on Windows and Mac, audio files are in Wave format playable on standard CD players.

Profitable Publishing for the Clueless®

The complete 3 CD set containing everything you need to know to generate, protect, and market your printed work. See full description for each item.

Disk 1 - Self-Publishing for the Clueless®
Disk 2 - Trademarks & Copyrights for the Clueless®
Disk 3 - Book Marketing for the Clueless®

Audio/Data CD+ ISBN 978-1-891440-51-9 **$59.95**

Project Management for the Clueless®

Zipped file containing a 1-hour audio seminar and all of the documents you'll need to organize, plan and budget most projects in three hours or less.

*eBook** ISBN 978-1-891440-95-3 **$9.95**

Self-Publishing for the Clueless®, 2nd Edition

Complete workbook on self-publishing, e-publishing, and a secured access to a 270 page manual with the latest information, resources, and insider tips about the world of e-publishing including how to be e-published in a week.

Workbook and Data CD+ ISBN 978-1-891440-99-1 **$34.95**

Trademarks & Copyrights for the Clueless®

Trademarks are the mark of your trade and copyrights address the laws allowing you the rights to make copies of your work. Contains printable forms and examples, explaining how to protect your works plus what material of others you can use without fear of legal problems.

Audio/Data CD+ ISBN 978-1-891440-30-4 **$24.95**
*eBook** ISBN 978-1-891440-92-2 **$9.95**

Venture Capital for the Clueless®

Venture Capital is available for all types of businesses and your ability to tap into it depends on your ability to write a business plan that sells you and your ideas to the people with the money. Includes explanations, samples, printable forms, and sources of venture funding.

Audio/Data CD+ ISBN 978-1-891440-31-1 **$24.95**
*eBook** ISBN 978-1-891440-93-9 **$9.95**

Whadda We Do NOW?™

Provides quick fixes for failing businesses. Learn how to quickly and easily figure out what's wrong so you can stop guessing and start implementing solutions. The information is practical, easy to understand, and readily implementable if you're serious about getting your business into a positive cash flow position-NOW!

Book ISBN 978-1-891440-66-3 **$29.95**

**eBooks contain a complete book in PDF for use on all Mac & Windows PCs, including desktops, notebooks/laptops, netbooks & tablets.*
+All data files are in PDF format, playable/viewable on Windows and Mac, audio files are in Wave format playable on standard CD players.

Training Kits

Become a Professional Organizer

If you'd like to get into the lucrative world of professional organizing, then everything you'll need to setup a business, get clients, and operate profitably is on the list below. Contains:

Books: How to Become a Clutterologist™, Clutterology® Getting Rid of Clutter and Getting Organized!; Fishin' with a Net™, How to Develop an Effective Web Site; Mechanics of Starting a Home-Based Business™, and Self-Publishing for the Clueless® 2nd edition

Plus DVD/CDs: Headline Creator™ Pro Suite, Marketing the One-Person Business™, and Highlights of Clutterology©

Kit ***$261.95***

Entrepreneurship

Do you dream of working for yourself? This kit includes everything for setting up a home-based business, getting organized, raising venture capital to fund the efforts, scheduling and managing your time, and ways to market your skills profitably.

Contains: Books: Mechanics of Starting a Home-Based Business™, Clutterology® Getting Rid of Clutter and Getting Organized!, Fishin' with a Net™, How to Develop an Effective Web Site, and Self-Publishing for the Clueless® 2nd edition.

Plus DVD/CDs: Marketing the One-Person Business™, Headline Creator™ Pro Suite, Professional Speaking for the Clueless®, and Venture Capital for the Clueless®.

Kit ***$169.95***

Invention Marketing

The material listed is endorsed by the SBA as "The only legitimate program for marketing inventions that we've ever seen." It explains how to organize and manage your invention process, protect them with patents, trademarks and copyrights, set up a home business, offer your ideas for sale, plus information for raising venture capital to fund your projects.

Contains: Books: How to Sell Your Inventions for Cash™ and Mechanics of Starting a Home-Based Business.

Plus DVD/CDs: How to Sell Your Inventions for Cash, Contracts and Agreements for Inventors, Venture Capital for the Clueless®, and Trademarks & Copyrights for the Clueless®.

Kit ***$161.95***

Professional Speaking

If you're interested in professional speaking, you'll find everything needed to get profitable bookings including places to get booked and instructions on how to do it! Includes over 1,700 pages of printable information, eight hours of video, four hours of audio, and hundreds of support resources.

Contains: Books: Mining the College Market, Mechanics of Starting A Home-Based Business, and Fishin' with a Net™, How to Develop an Effective Web Site, Self-Publishing for the Clueless® 2nd edition, and Business Training for the Clueless® (plus CDs),

Plus DVD/CDs: Headline Creator™ Pro Suite, Marketing the One-Person Business, Professional Speaking for the Clueless®.

Kit ***$461.95***

Raising Venture Capital

Looking for money or backing for a new idea or enterprise? Confused about what to do and how to approach investors? We've assembled the materials needed to establish proprietary rights to your innovations; plan, budget, organize and schedule your project; prepare a business plan and shop it to people with investment capital. Contains:

Books: Clutterology® Getting Rid of Clutter and Getting Organized! and Mechanics of Starting a Home-Based Business, Fishin' with a Net™, How to Develop an Effective Web Site, and Self-Publishing for the Clueless® 2nd Edition.

Plus DVD/CDs: Trademarks & Copyrights for the Clueless®, Venture Capital for the Clueless®, Headline Creator™ Pro Suite, and Highlights of Clutterology©.

Kit ***$191.95***

**eBooks contain a complete book in PDF for use on all Mac & Windows PCs, including desktops, notebooks/laptops, netbooks & tablets.*
+All data files are in PDF format, playable/viewable on Windows and Mac, audio files are in Wave format playable on standard CD players.

Self-Publishing

You CAN have a book ready to sell in 30 days with these practical products. . . guaranteed! Contains everything needed including pre-configured scheduling and budgeting charts to get your own project finished in record time with a minimal amount of expense and hassles.

Contains: Books: Mechanics of Starting A Home-Based Business, Fishin' with a Net™, How to Develop an Effective Web Site, and Self-Publishing for the Clueless® 2nd edition.

Plus DVD/CDs: Trademarks & Copyrights for the Clueless®, Book Marketing for the Clueless®, and Headline Creator™ Pro Suite.

Kit ***$261.95***

Web Site Development Software Suite

No software you use to create your Web site will contain everything you need, so we've assembled the "stuff they left out." This suite of software and resources will help make your Web site work efficiently and get the response it deserves. These products have a proven track record in the world of Web design, marketing and advertising.

Contains: DVD/CDs: AV Toolkit; Photo Manipulator Toolkit, Headline Creator™ Pro Suite, Trademarks & Copyrights for the Clueless®, and E-Commerce for the Clueless®.

Kit ***$100.00***

**eBooks contain a complete book in PDF for use on all Mac & Windows PCs, including desktops, notebooks/laptops, netbooks & tablets.*
+All data files are in PDF format, playable/viewable on Windows and Mac, audio files are in Wave format playable on standard CD players.

Rounds Miller and Associates

6318 Ridgepath Court
Rancho Palos Verdes, CA 90275-3248
310.544.9502
www.RoundsMiller.com

Business, Technology and Organizing Training Specialists

ORDER FORM

ITEM (See Catalog for Full Description)	Format	Qty.	Price
Audio/Video Toolkit	Audio/Data CD $24.95		
Book Marketing for the Clueless®	Audio/Data CD $24.95		
	eBook - Online $9.95	PURCHASE ONLINE	
Business Training for the Clueless®	Book & Data CD $69.95		
	eBook - Online $9.95	PURCHASE ONLINE	
Cash From Your Clutter	eBook - Online $9.95	PURCHASE ONLINE	
The Clutter Bug Attacks Christmas Clutter™	Book $15.95		
	eBook - Online $9.95	PURCHASE ONLINE	
The Clutter Bug Attacks Junk Mail, Spam and Telemarketers™	eBook - Online $9.95	PURCHASE ONLINE	
The Clutter Bug Investigates Coupons, Discounts, and Deals™	Book $24.95		
	eBook - Online $9.95	PURCHASE ONLINE	
Clutterology® Getting Rid of Clutter and Getting Organized!	Book $34.95		
	eBook - Online $9.95	PURCHASE ONLINE	
Highlights of Clutterology®	Audio CD $19.95		
How to Become a Clutterologist™	Book $34.95		
	eBook - Online $9.95	PURCHASE ONLINE	
Consulting for the Clueless®	Book/Software DVD $39.95		
Contracts and Agreements for Inventors	Data CD $19.95		
E-Commerce for the Clueless®	Data CD $24.95		
Fishin' With A Net	Book $24.95		
Headline Creator™ Pro Suite	Software CD $39.95		
How To Sell Your Inventions for Cash	Book $24.95		
	eBook - Online $9.95	PURCHASE ONLINE	
	Audio CD $39.95		
Intellectual Property Protection for the Clueless®	MP3 Audio/Data CD $59.95		
Marketing the One-Person Business	eBook - Online $9.95	PURCHASE ONLINE	
Mechanics of Starting a Home-Based Business	Book $34.95		
	eBook - Online $9.95	PURCHASE ONLINE	
Photo Manipulator Software	Software CD $19.95		
Professional Speaking for the Clueless®	Audio/Data CD $24.95		
	eBook - Online $9.95	PURCHASE ONLINE	
Profitable Publishing for the Clueless®	3 Audio/Data CDs $59.95		
Project Management for the Clueless®	eBook - Online $9.95	PURCHASE ONLINE	
Self-Publishing for the Clueless®	Workbook & Data CD $34.95		
Trademarks & Copyrights for the Clueless®	Audio/Data CD $24.95		
	eBook - Online $9.95	PURCHASE ONLINE	

Page 1 Sub-Total	

Venture Capital for the Clueless®	*Audio/Data CD $24.95*		
	eBook - Online $9.95	PURCHASE ONLINE	
Whadda We Do NOW?™	*Book $29.95*		
TRAINING KITS			
Become a Professional Organizer	*Kit $261.95*		
Entrepreneurship	*Kit $169.95*		
Invention Marketing	*Kit $161.95*		
Professional Speaking	*Kit $461.95*		
Raising Venture Capital	*Kit $191.95*		
Self-Publishing	*Kit $261.95*		
Web Site Development Software Suite	*Kit $100.00*		

E-Books are available for purchase and instant download online at www.RMACart.com/E-Books.htm

Amount from Page 1	
Sub-Total	
in CA add Sales Tax 9.75%	
Shipping	**$2.95**
Total	

Thank You!

Name (please print) ____________________

Mailing Address: ____________________

City, State, ZIP: ____________________

Tel: ____________________ e-Mail: ____________________

I authorize Rounds, Miller and Associates to charge my credit card for the items listed above

VISA | American Express Cards | MasterCard | Discover Novus

Credit Card Number: ____________________ exp. __________ CSS__________

Signature: ____________________ Date: __________

To Order By Mail:

Send completed order form and check payable to Rounds, Miller and Associates to

6318 Ridgepath Court

Rancho Palos Verdes, CA 90275

View our entire product line at www.RoundsMiller.com

Printed in Dunstable, United Kingdom

91335489R00098